CREATIONISM:
DESIGN ERRORS AND CROSS-PURPOSES

This book is published by The Lindsey Press on behalf of the
General Assembly of Unitarian and Free Christian Churches.

Unitarianism is a progressive religion whose historical roots lie in
liberal Christianity. Its congregations have evolved over the past
400 years into multi-faith communities which include humanists,
atheists, and agnostics.

Unitarianism draws on insights from many religious, philosophical,
and scientific traditions. It emphasises the primacy of individual
reason, conscience, and experience in matters of faith, and it is strongly
committed to social action and human rights. Unitarians have been in
the vanguard of many social reforms, campaigning most recently for full
equality for same-sex marriage, and for the right of incurably ill people to
seek assistance to end their lives.

www.unitarian.org.uk

CREATIONISM:
DESIGN ERRORS AND CROSS-PURPOSES

Graham Richards

The Lindsey Press
London

Published by the Lindsey Press
on behalf of the General Assembly of Unitarian
and Free Christian Churches
Essex Hall, 1–6 Essex Street, London WC2R 3HY, UK

ISBN 978-0-85319-084-4

Designed and typeset by Garth Stewart, Oxford
Front cover image by Shutterstock / Vladimir Arndt

Printed and bound in the United Kingdom by
Lightning Source, Milton Keynes

For Hilary Brown, Geoff Bunn, Lynn and Graham Clarke,
Carl O'Dea and Dan Huckfield, Jane Munroe, Jenny and Colin Roberts,
and Viv and Nigel Suckling

Also in memory of Geoff Midgley and Leslie Wyner

And what a releasing moment of freedom it is when, having been hampered by being told we *ought* to believe some improbable thesis because the church teaches it, or the Bible states it, we throw it away and discard it forever for the nonsense it has really always been!

Leslie D. Weatherhead (1965) *The Christian Agnostic*, p.31

Contents

Foreword

My somewhat lame credentials for writing this foreword to Graham Richards' splendid meditation on Creationism, Intelligent Design (ID), and science are that I have a PhD in Molecular Biology (University of Wisconsin 1969, when Molecular Biology meant something rather different from its present meaning), and that I have written articles and given talks on Creationism and ID, usually to Unitarian audiences, where the take-home lesson was that ID is not science, and must be kept out of school science lessons. I proposed a motion to the same effect at the General Assembly of Unitarian and Free Christian Churches in 2006. The motion was passed.

Furthermore, during the flurry of evolutionary excitement in 2009, the bicentenary of Darwin's birth, I gave several talks to more general audiences. One of these, quite without any conscious intention, took place at Wakefield's Westgate Chapel on 24th November, the 150th anniversary of the publication of *On the Origin of Species*! When this coincidence finally dawned on me, I wondered – not for the first time – about the existence of guardian angels!

During 2009 we Unitarians, of course, tried to suggest that the Unitarian influence on Darwin, from both his maternal and paternal ancestry, as well as from his wife, was crucial in nurturing his scepticism concerning the orthodoxy of the day and thus led to his penchant for asking the large general questions that culminated in the Theory of Evolution through Natural Selection. For a discussion of Unitarian influences on Darwin, see Cliff Reed's cogent little book *'Till The Peoples All Are One'* (Lindsey Press, 2011).

But now, having read *Creationism: Design Errors and Cross-Purposes* twice, I realize that my approach to the subject of Creationism and ID was too narrow. It usually took the following pattern. After emphasising the importance of Creationism to one strand of US politics, I would outline the history of the several attempts to get it included in the national high-school science curriculum – attempts that failed as they broke against the rock of the separation between church and state. ID was introduced in response to these failures, as a stratagem to convince the courts that it was science, not religion, because the Designer was not specified as the biblical God. In the first legal test of ID the court saw through the stratagem. Following on from this, I would criticize ID as being merely a collection of perceived flaws, gaps, or impossibilities in the Theory of Evolution, rather than a testable scientific hypothesis that one could find evidence for or against. How could one obtain reproducible evidence for the existence of a designer?

I would usually finish by pointing out that the exciting newly deciphered genome of humans (and virtually all other organisms) is littered with pseudogenes, which are sequences in the DNA that are clearly inactivated genes. Would an intelligent designer put in genes that do not work? One of those pseudogenes present in primates is an inactive remnant of a gene that allows other animals to make their own vitamin C and is the reason why humans, when deprived of the vitamin in their diet, die horribly of scurvy.

The plethora of DNA sequences that we now know is exactly what we would expect from the Theory of Evolution but is almost impossible to explain as the work of an Intelligent Designer – unless It is an Intelligent Deceiver. Michael Behe, a proponent of ID, attempts to account for pseudogenes and other anomalies or apparent errors of design as follows: '... features that strike us as odd in a design might have been placed there by the designer for a reason – for artistic reasons, for variety, to show off, for some as-yet-undetected practical purpose, or for some unguessable reason – or they might not' (Michael Behe, *Darwin's Black Box*, p. 223). The Theory of Evolution, on the other hand, is strengthened by these odd features, because evolution has to start from what is already present, warts and all. And had the genome sequences of mice and humans turned out to be more similar to each other than those of chimps and humans, the Theory of Evolution would have been dead in the water! (See the discussion of Falsifiability in Chapter 4 of this book.) So our conclusion is that the Theory of Evolution has passed another stringent test.

In this book, Professor Graham Richards takes a much broader, more philosophical, more compassionate, and less strident position than I used to take. He presents the most lucid and accessible explanation of the scientific adventure that I have ever read. He justifiably criticizes scientists as well as Creationists, whose motives and culture he made me understand so that I sometimes felt sad for them as I read. There is a wonderful chapter on the Bible and why it has endured so well and stayed relevant, in contrast to the scientific literature which is relevant for only an instant. I learned about Aristotle's four types of cause, and the three anxieties identified by Paul Tillich, which seemed so right, and I found the three appendices very informative, even though I was already familiar with *The Book of Genesis*. In fact I think I learned important, relevant things from every chapter. And the approach is consistently witty and engaging, never dull.

I believe this book will educate and appeal to non-scientists as well as scientists, to humanists as well as the religious. Indeed, I intend to recommend it to my book group, only one other of whom is a scientist! I believe they will find it an engrossing read.

Simon Hardy BA (Cantab), PhD (Wisconsin)
Member of St Saviourgate Unitarian Chapel, York, UK

Preface

I should explain a little about how this book originated. It was first drafted fairly intensively in an eight-week period as long ago as April–June 2006, and it has been intermittently expanding ever since. It was triggered by a blend of two frustrations: the first with developments in British education at the time, such as the City Technology College in Gateshead, sponsored by ardent Creationist Sir Peter Vardy, which seemed to be endowing Creationism with some kind of academic respectability; and the other with the terms in which the Creationism controversy was being fought – by both sides – when viewed from the perspective of an historian of Psychology concerned with the complex historical relationships between Psychology and religion. The first of these frustrations, was, it transpired, a little overstated, since the college was in fact teaching the standard National Curriculum, and Vardy's own approach to education (though not his own Creationist beliefs) had been somewhat misrepresented in the press. In educational contexts the issue, although still of some concern, now figures less prominently in public debate, as educational policy in general becomes the target of heated discussion; but even so it lurks at the grass-roots level. Creationist science teachers are far from unknown in UK secondary schools.

However, the disturbing nature of the Creationism debate itself has, if anything, worsened, and since that initial effort a number of further contradictions and problems with the positions of Creationism and Intelligent Design (ID henceforth) have dawned on me and have been incorporated in this text. The anti-Creationism/ID camp has, however, largely continued to argue along lines which have long proved ineffective. Rather than concluding, in proper scientific spirit, 'we must be doing something wrong here: empirical scientific arguments are just not working', they blame their failure simply on the intellectual obtuseness and ignorance of their opponents – and try all over again. Accordingly I have attempted both to propound arguments of a different kind and to diagnose the roots of the Creationist/ID resistance to what seem to be quite clear-cut scientific refutations. In a way it has been this challenge and the frequent emergence of new angles on the topic which have kept me going.

Nearly at the outset I was encouraged by a conversation with John White, recently retired Professor of Philosophy of Education at London University's Institute of Education, in which he keenly stressed the need for such a work. The controversy itself is still not quite as culturally prominent in Britain or mainland Europe as it is in the United States, but pro-Creationist and ID views are still being covertly smuggled into the classroom via the current government's encouragement of 'faith schools', although rarely in science lessons themselves, constrained as they are by the terms of the National Curriculum. I hope that what follows makes a modest contribution to countering this tendency.

One consequence of seven years of revision and expansion, as described above, is that the text has become increasingly digressive in style on occasion, although the relevance of such digressions to the main arguments has, I hope, always been made clear.

The final stage in the book's production came when my friend Professor Hilary Brown suggested that I approach the Lindsey Press with a view to publication. The airing of any more new thoughts on the topic will have to be postponed!

Graham Richards
Tunbridge Wells, January 2014

Acknowledgements

I would like first to thank the team at the Lindsey Press, representing the General Assembly of Unitarian and Free Christian Churches, who decided to adopt the book, and especially their anonymous reviewer, whose enthusiasm finally tipped the scales in its favour.

For their support and feedback on earlier drafts I am forever grateful to Hilary Brown, Geoff Bunn, Graham Clarke, Paul Dunn, Vincent Hevern, Dan Huckfield, Arthur McCalla, Mary Midgley, Roy Perkins, Robert Rieber, Colin Roberts, Jenny Roberts, Sonu Shamdasani, and John White, plus several anonymous earlier reviewers. Maura's loving support has, as ever, been unstinting.

Introduction

This book aims to introduce the reader to the controversy over Creationism and Intelligent Design (ID), demonstrating in the process numerous central flaws in the notion of ID as a 'scientific' position ('Creation Science'), and more explicitly the flaws in religious fundamentalist 'Creationism' (although not all Christian 'fundamentalists' are in fact Creationists). These flaws are, I believe, fatal to both positions. In Britain, this issue continues to rumble on, while in the United States it remains a fiercely fought cultural war-zone, and (to mix metaphors) the ripples from it are lapping at European shores on a more wave-like scale. Whether the tide is actually ebbing or flowing is, however, difficult to judge with any confidence.

Having said that, I should make it clear that anyone expecting to read here a passionate, atheistic, 'materialist' assault on religion in general, or even on those espousing these specific doctrines, will be disappointed. My academic background is a mixture of Psychology (particularly its history) and Philosophy, while at one time I became deeply involved with the question of human evolution, which yielded my first published book in 1987. Obviously the extent of my professional expertise varies fairly widely across the numerous different disciplines on which I have had to touch, but I have striven to ensure that I have not misrepresented or misunderstood their bearings on the issues in question. Some readers claim to detect a 'Protestant' bias in my position. Any such bias has been unwitting, although I concede that the Catholic view of the relationship between religion and science has its own unique features, which I have not addressed. How far these might affect the acceptability of my general reading of the Creationism issue to Catholic readers I cannot say.

While lacking personal religious affiliations, I am not insensitive to the convictions of those who do avow religious faith, but – to make my position clear at the outset – I believe that reversion to ancestral religious creeds, rooted in cultural contexts profoundly different in character from *any* in which people now live, is mistaken. If the needs commonly called 'spiritual' are to find any satisfaction, we must move forwards not backwards, jettisoning vast swathes of traditional religious doctrine, not least the claims to exclusive possession of the truth made by Christianity and Islam. The solution cannot lie in 'science' as such, since – as we shall see – that is not its job. It is not even the job of the discipline of psychology, despite having a sub-discipline called 'the psychology of religion'. It is, rather, a collective task in which all can bring their insights to the debating forum. For present purposes, however, that broader issue is by the by – or nearly so. Some apparent deviations from the main theme are essential if the topic is to be tackled in anything approaching an adequate fashion. In particular, some criticisms of the position of contemporary physical science could not be avoided.

The book was provoked by the current resurgence of fundamentalist religious thinking across many faiths, of which all are aware, and what I believe to be a dangerously regressive rise in its respectability. Fundamentalist religion, especially in the United States, is making increasing inroads into education, and the doctrines of ID, Creationism, and anti-evolutionism are often a central feature of this trend. In the United States and, to some extent, in Britain there is pressure (as yet successfully resisted) for these ideas to be taught as part of the science curriculum as genuine alternative 'theories'. Many religious believers, of course, share my disquiet and unease.

What follows is therefore an attempt to identify a number of errors (some quite straightforward, others more subtle) in the thinking that underlies such doctrines, and to offer a number of observations, variously psychological, philosophical, historical, and linguistic, regarding what might lie behind them. I cannot promise to avoid the odd sarcastic or contemptuous flourish entirely, particularly when faced with ideas which strike me as especially inane; but I am, I stress, unconcerned with engaging in polemical and confrontational rhetoric. Indeed, it became clear to me that the current situation is actually something of a *folie à deux*. In writing the book I was not, then, entering the lists as a passionate defender of the physical sciences, but as some kind of historical-cum-philosophical psychologist. As such, my goal was not really outright victory, even though I naturally hoped that my criticisms of Creationism and ID would be convincing. Rather, I sought a philosophical and psychological diagnosis of the controversy which, it turned out, did not leave the opponents of Creationism with an entirely clean bill of psychological health either. In other words, both camps, not just the religious protagonists, needed to rethink the roots and nature of the controversy. The reader must judge whether the goal that I set myself has been reached.

It is, finally, worth pondering what 'outright victory' would actually look like. Does anybody really believe that the other side can be manoeuvred into unconditional surrender? Or intellectually and morally annihilated in some fashion? Is the defeated party expected to flood the media with mass apologies and recantations? Envisaging outright victory in this, as in many other contexts, is simply fanciful.

1 What is creationism?

Creationism is one central doctrine of much contemporary fundamentalist Christianity, along with other notions such as the widespread expectation of the Second Coming (often called 'Pre-Millennialism'). While Creationism is not confined to Christian fundamentalism, shared as it is by Jewish and Islamic fundamentalists, the Christian version dominates the debate, and it is with this that I shall be concerned here.[1] It is also important to see the contemporary controversy in a much wider historical context, and we will be doing this intermittently throughout what follows, particularly in Chapters 5 and 6. At this point it only needs to be emphasized that – as Karen Armstrong (2000) explains in her book *The Battle for God. Fundamentalism in Christianity, Judaism and Islam* – fundamentalism flourishes most when communities find themselves exposed to radical social changes and upheavals – especially, in recent times, those associated with the impact of 'modernism'. Such events are experienced as threatening their cultural identity.[2] This has been particularly evident over the last two centuries, as successive waves of 'modernism' swept first across western Europe and thence successively into the northern United States, across the European empires (including the Middle East) and, by the latter part of the twentieth century, to the rest of the United States, Japan, India, China, and much of Africa and South America. But what exactly is Christian 'Creationism'?[3]

Christian Creationism

At its simplest, Christian Creationism is the doctrine that the universe was created by God in the way described in the biblical *Book of Genesis*, Chapters 1–8. In this account it seems that there were three stages in the creation of the present world. First, a six-day phase culminating in the creation of Adam and Eve. Secondly, the Fall, following Adam and Eve's eating of the forbidden fruit from the Tree of Knowledge which resulted in their expulsion from the Garden of Eden. Thirdly, and around 2,000

4

years later, came Noah's Flood, which destroyed all land- and air-dwelling life forms except those on the Ark. While the information provided in the text is scanty – and actually ambiguous, since two versions of the first phase are offered – it is clear that both the Fall and the Flood radically altered the world as initially created. Many Creationists believe that such phenomena as fossils and the Grand Canyon were produced by the Flood, and that dinosaurs co-existed with humans until then. Further thoughts on this account are given in Appendix B.

Contemporary Creationism, as advocated primarily in the United States, takes two major forms, however. The first, often referred to as the 'Young Earth' position, which we may call the 'Strong' version, accepts that a literal reading of the *Genesis* account requires that the first stage took place around 6,000 years ago. In this view the entire modern scientific cosmology is erroneous, from the size of the universe and distances between stars to the nature of fossils and, of course, the origin and evolution of life on Earth. In the United States this doctrine is also often one component of the broader fundamentalist Pre-Millennialist world view mentioned above, which believes that the end of the world, as foretold in the *Book of Revelation*, is imminent and that its Apocalyptic events are – or will very shortly be – in the course of unfolding.[4]

The second version claims, if often unconvincingly, to be independent of religious belief, but argues that it is nonetheless necessary to postulate the existence of a 'designer' in order to explain the complexity of life forms and the basic laws of physics which underpin the universe. This is known as the 'Intelligent Design' (ID) position. Its exponents are generally less committed to the 6,000-year time-span, but are particularly hostile to mainstream evolutionary thought. It should, however, be noted that within the 'Strong' camp there have been differences of opinion, often leading to splits. In particular there have been those who argued for what is known as a 'Gap' theory, in which the events of *Genesis* replaced a pre-existing world of unknown duration, geological traces of which have survived. This 'progressive creationism', as Ronald Numbers terms it, increasingly merged into 'Scientific Creationism' during the twentieth century.[5]

Emergence of Creationism in the seventeenth century

Both strands mentioned above are traceable in something resembling their present character to the seventeenth century. Before that, the *Genesis* account of the creation was universally accepted in Christian, Jewish, and Islamic traditions; but, in the absence of any astronomical and geological knowledge which might shed doubt on its literal truth, this was basically a point of religious doctrine which few were interested in contesting. More than that: it was a fertile source of religious artistic imagery, providing a quite literally iconic opening to depictions of the Christian cosmology, with huge scope for creative exploration and celebration. By the end of the sixteenth century fundamental challenges to traditional understanding of the physical universe were mounting, notably from astronomers increasingly inclined to accept Copernicus's heliocentric theory that the Earth went round the Sun, a theory which displaced the Earth from its central position in the universe. Encounters with the peoples of the New World and elsewhere were another source of unease, for their very existence, variety of physical appearance, and diverse cultural characters were difficult to reconcile with the dogma that all peoples were descended from the sons of Noah. (Sir Walter Raleigh made one valiant attempt at doing so in his 1614 *The Historie of the World*.)

In this proto-scientific climate, in which quantification and measurement were beginning to assume a central place, one response among many was to scrutinize the biblical account more closely. Thus, it is typically claimed, in the 1650 *Annales Veteris Testamenti*, co-authored with James Flesher and Laurence Sadler, James Ussher, Archbishop of Armagh, (1581–1656) published his calculation that God had created the world at 9 a.m. (time-zone unspecified) on Sunday 23 October 4004 BCE. Actually, around 140 different dates calculated on the biblical information had been proposed by the mid-nineteenth century, ranging from 3616 to 6984 BCE, but Ussher's date won out in the evolution of Creationist thinking. Yet, ironically and reflexively reinforcing the unreliability of dating, the precise dating and details of this very episode – as well as the calculations – have themselves subsequently become confused, misquoted, and mythologized. To quote a current web-source: 'In 1642,

Dr John Lightfoot wrote that man was created at 9:00 a.m., and in 1644 he wrote that the world was created on Sunday, September 12, 3928. In 1650, the Irish Archbishop, James Ussher, published his computations that the world was created on Sunday, October 23rd, 4004, beginning at sunset of the 22nd. Both these dates are widely misquoted.'[6] So Lightfoot's hour and Ussher's date have been conflated in the collective creationist memory. 'Strong Creationism' in the modern sense may, a little simplistically perhaps, be said to date from Lightfoot's and/or Ussher's arrival at this strangely precise moment in time. The victory of the Ussher/Lightfoot dating was largely due to Ussher's, in particular, being incorporated into the annotations and cross-references in a 1701 edition of the King James Authorized Version of the Bible, after which it tended to be retained until the late nineteenth century and was perpetuated in the influential *Scofield Reference Bible* (1909) (see below, p.73).[7]

While its origins, character, and motives are a matter of continual debate by historians of science, everyone agrees that something worthy of the name 'The Scientific Revolution' occurred during the seventeenth century and was already in progress at the time when Ussher was writing. By the early 1700s the 'Newtonian' world view had won wide support among 'Natural Philosophers' across Europe, being challenged primarily by the 'mechanists', who could not accept gravity's 'action at a distance'.[8] This was popularly understood as having revealed the mathematical laws governing gravity and the motions of the universe. Isaac Newton, it seemed, had unveiled the marvellously harmonic character of Nature itself, and in so doing confirmed and deepened belief in the wisdom of its divine designer. Newton's own religious motives are, though rather complex, generally accepted, and similar motives were prominent among his generation of natural philosophers. One influential work from this period was John Ray's *The Wisdom of God Manifested in the Works of Creation* (1691), which went into numerous enlarged editions over the following century. (Ray was a prominent figure of the Scientific Revolution who had initiated a revolutionary empirical approach to animal and plant classification.) As the title indicates, this work was concerned with expounding the case for divine design, and it is claimed by Ray's leading biographer, C.E. Raven (1942), that William Paley, of whom more later, heavily plagiarized it. In one sense the very notion that the physical universe was rationally ordered and thus comprehensible was itself a Creationist premise underlying the birth of modern science. One widely held aspiration was that through such knowledge it would be possible to redeem ourselves – and the world – from the ignorance and other consequences of the Fall, thus returning to a 'Prelapsarian' state.

Design and 'Natural Theology'

However, not all eighteenth-century 'Natural Philosophers' (the word 'scientist' did not become current until the 1820s) agreed with this religious perspective on Newton's achievement. In France especially, the rapid growth of scientific knowledge fostered materialist and atheistic philosophical positions, while the aforementioned 'mechanistic' view lingered through much of the century. Even so, in Britain the assumption prevailed that this thriving expansion of knowledge reinforced what was being called 'Natural Theology', constituting a reading of the mind of God as revealed in the Book of Nature, and complementing the spiritual knowledge divinely revealed in the Bible. This being so, the notion of Intelligent Design was implicit in much of what passed for 'scientific' knowledge throughout the 1700s. Nevertheless, since the existence of God was, with some exceptions, a generally unchallenged assumption, the Argument from Design itself was rarely fully re-articulated following the publication of John Ray's book. Science revealed God more fully, rather than proving His existence. It is interesting that it was, at this time, the *harmony* rather than the *complexity* of the Newtonian universe that was felt to signify God's presence so powerfully. 'Harmony' is fundamentally an aesthetic rather than a physical property – and preference for this term suggests that awe of the universe as an artistic achievement initially trumped admiration of it as a feat of mechanical engineering.

All the same, there were accumulating difficulties, particularly regarding time-scales.[9] Although still at an early stage of its development, geology was, by 1800, already yielding numerous findings, theories, and hypotheses hard to reconcile with a literal reading of *Genesis*. Rudimentary calculations of the rates at which some sedimentary rocks were deposited suggested that stratified beds hundreds of feet thick could not be fitted into an Ussher-like time-span. And while there was plenty of evidence for floods and past inundations of what is now dry land, these appeared far more complicated than a one-off flood such as Noah's could explain. Then there were the fossils, the true nature and implications of which had only begun to be understood during the mid-eighteenth century. It was increasingly difficult to explain the fossil evidence, such as their distribution through different strata, as the product of a single inundation. Astronomy too was beginning to grasp the immense distances and scale of the universe. Even if as yet reluctant to abandon ID, more and more geologists and astronomers became disinclined to take the *Genesis* story literally. Such doubts were further reinforced on a quite different front. A new brand of critical biblical scholarship was being pioneered in Germany which strove to set biblical texts in their full historical context and was additionally informed by developments in linguistics.[10] This involved unravelling the various authorships, revisions, amendments, redactions, and censoring which these texts had undergone in the distant past. French scientists such as the astronomer Pierre-Simon Laplace (1749–1827) were meanwhile increasingly adopting the materialist and atheistic positions. For the first time it began to dawn upon various thinkers that for scientific purposes God might prove to be dispensable: Laplace, presenting his discourse on secular variations of the orbits of Saturn and Jupiter to Emperor Napoleon I, was asked '*Mais où est Dieu dans tous cela?*' ('Where is God in all of this?'); to which Laplace is said to have replied: '*Je n'avais pas besoin de cette hypothèse-là*' ('I had no need of that hypothesis').

Thus it was that the theologian and philosopher William Paley (1743–1805) came to publish his *Natural Theology* (1800), the most influential and frequently cited statement of the Argument from Design. He famously began with the example of the need to infer the existence of a watchmaker if one chanced upon a watch lying on the ground, since obviously nothing so complex could have come about by chance. It therefore seemed to follow that far more complicated natural phenomena, especially the anatomy and physiology of living organisms, must have similarly been designed. At first glance this argument appears highly plausible and even today it remains at the core of Creationist and ID thinking. Two crucial but often overlooked points have to be made here, requiring a slight digression.

Digression: two key points

First, the Argument from Design is, in Paley's widely accepted version, an argument for the existence of God (*from* Design *to* God). It is *not* being offered as an explanatory scientific hypothesis – the designed character of the natural world is, once pointed out, perfectly obvious and uncontestable. It has, however, no consequences at all for scientific research methods and theorising on specific topics. This is actually the very reverse of the way in which it is used in current ID thinking, in which it is proposed as being scientifically necessary. This is because, the obviousness of design having long evaporated, the Creationist task became to hunt down – to 'scientifically discover' – an irrefutable example of it: a problem which Paley simply did not have to confront.

Secondly, what gave the concept so much of its power was that the whole purpose of the Creation came to be seen as to awe, educate, and nurture God's culminating achievement – Mankind. As the titles of the Bridgewater Treatises indicate (see below), the entire cosmos, especially earthly Nature, was 'adapted' to human needs. Again, this interpretation of the natural world, once so obviously true, simply fails to convince any longer.

There is then a radical difference between the Argument from Design as conceived by Paley and his contemporaries and present-day ID theories. I will return to this issue at various later points, but it is vital to grasp at the outset the difference between the uses of the Argument from Design made by Paley and those made by the modern proponents of ID. In the next chapter we will address more fully the argument from complexity to design advocated by Paley, which remains central to ID thinking.

The nineteenth-century decline of Natural Theology

Paley's argument flourished in the early nineteenth century, and a series of essays, known (after the Earl of Bridgewater, who left the bequest which funded them) as *The Bridgewater Treatises*, provided its most solid articulation. These had titles such as *On the Adaptation of External Nature to the Physical Condition of Man* (by John Kidd, 1833) and *The Hand: Its Mechanism and Vital Endowments as Evincing Design* (1832, by the eminent surgeon and anatomist Charles Bell). William Whewell, one of the greatest of the early nineteenth-century philosophers of science (often credited with coining the very word 'scientist'), contributed *Astronomy and General Physics Considered with Reference to Natural Theology* (1833). But even before Paley, the Scottish philosopher David Hume had produced a powerful critique of the Argument from Design, published after his death as *Dialogues Concerning Natural Religion* (1779). By the 1840s Paley's simple version of the argument was already under strain and yielding to rather more sophisticated arguments for 'Theism', the notion that Mind or God must be present in Nature. These did not rely on the design analogy as such. (We will return to Theism in Chapter 7.)

It was during the latter part of the eighteenth century and the early nineteenth century that the two strands previously identified really began to diverge. In ever greater numbers, scientific thinkers and the intelligentsia in general no longer felt that their faith demanded a literal belief in the six-day Creation story, increasingly accepting time-scales extending into five, six, or even seven figures (still, of course, much shorter than present estimates). At the same time, suspicions were growing that the story of Noah's Flood rather oversimplified actual events.

Finding these trends hard to accept, the more evangelical and fundamentalist churches and denominations began digging their heels in, expending great effort in trying to counter such arguments and defend the literal truth of the *Genesis* account. Far from being a return to a long-obscured tradition of biblical understanding, strict thoroughgoing literalism was in fact a largely nineteenth-century development. And now, as we shall see in Chapter 6, the European and North American stories also began to diverge. In Britain and mainland Europe the 'Strong'

Creationist position became ever more marginalized as the nineteenth century progressed. By its close, following the triumph of evolutionary ideas (including non-Darwinian versions) after 1860, fundamental shifts in geological understanding, and a tide of new fossil discoveries, only a handful of minor sects were still adhering to it. (Evolutionary ideas had been around since the end of the previous century, but it was the publication in 1859 of Charles Darwin's *On the Origin of Species by Means of Natural Selection* that proved scientifically crucial.) A common anti-evolution argument at this time was the so-called 'God of the Gaps' theory, which claimed that divine explanations were necessary to explain, for example, gaps in the fossil record. This ploy was soon abandoned as accumulating evidence kept plugging such gaps. In the United States, by contrast, its decline was far less dramatic and widespread, and in fact was effectively reversed to some extent between 1880 and the 1920s.[11]

Creationism in the early twentieth century

During the twentieth century British and European religious hostility to the theory (or, more accurately, theories) of evolution was fairly muted, although many Christians of all varieties continued to feel that it was, if not wrong, at any rate insufficient. Rather moderate ID positions, sometimes of the 'gap' variety, were not uncommon but were rarely passionately advocated beyond more fervent fundamentalist circles, and certainly no direct confrontation with science appears to have been sought.[12] During the 1950s, I recall, a series of pro-ID 'Fact and Faith' films was circulated for showing by the Christian Union, a mainstream organization popular among Christian secondary-school students, but this raised little controversy and our science teachers paid it scant attention. Europe had other more urgent things on its collective mind throughout most of the century. In the United States, however, anti-evolutionists remained ever restless, and 'Darwinism' was frequently cast by Creationists as directly responsible for the catastrophes sweeping Europe, from the Great War to the rise of Nazi Germany.[13] The most famous episode was the 1925 Scopes Trial in Tennessee (sometimes called 'the monkey trial'), at which a teacher was arraigned for defying a local law banning the teaching of evolution in schools. This episode has proved persistently fascinating and was the subject of the play *Inherit the Wind* by Jerome Lawrence and Robert E. Lee, first performed in 1955 and subsequently filmed by Stanley Kramer (1960), with a TV-rewrite broadcast on NBC in 1988. Although the teacher was found guilty, the leading pro-Creationist witness, William Jennings Bryan, was so severely humiliated by defence counsel Clarence Darrow that Creationism's national progress ground to a halt.[14] At the same time, and into the 1930s, a geologist, George McReady Price, was publishing popular Creationist geology textbooks of a kind which had disappeared in Britain by the end of the 1880s. Price first broke cover with a 1906 pamphlet entitled 'Illogical Geology' and published the full version of his views as *The Fundamentals of Geology* in 1913; a textbook version entitled *The New Geology* followed in 1923, and a revised edition of the first version, re-titled *Evolutionary Geology and the New Catastrophism*, appeared in 1926.[15] Probably the last major British Creationist work of this kind had been the Revd. Samuel Kinns' *Moses and Geology* of 1882, the most fully elaborated of a long-standing genre of books mapping the new geology on to the biblical account. Before the first World War the major target of much US Creationism and fundamentalist Christianity was actually the 'higher criticism' of the Bible, initiated by Schleiermacher, rather than evolutionary theory as such, which was, at the time, apparently in some scientific difficulties. Only in the 1920s did anti-evolutionism re-assume central importance.

Re-emergence of Creationism since the 1960s

Between the 1950s and early 1970s the triumphantly optimistic character of post-war science in the United States created a climate in which, despite the wider revival of Billy Graham-style evangelical fundamentalism, Creationism and ID as such held little appeal beyond the already converted in their existing heartlands of the Mid-West and Deep South. The first stirrings of a major revival came with the 1961 foundation of the Creation Research Institute by Henry M. Morris, who the previous year had published *The Genesis Flood. The Biblical Record and Its Scientific Implications*, co-authored with John C. Whitcomb Jr (Morris being the second author). This organization replaced the earlier 'American Scientific Affiliation' group, which Morris and his associates felt had gone soft on evolution.[16] The times were not yet ripe for a breakthrough, however. Sensational fossil discoveries such as that of the early hominid 'Lucy' in the Afar Triangle region of Ethiopia (1974) were widely publicized, coupled with breakthroughs in using mitochondrial DNA-based techniques for dating the split between human and chimpanzee lineages (at c. 6 million years ago, far more recent than earlier estimates of 14 million years, thus revolutionizing theories about the course of early hominid evolution). Such discoveries reinforced the popular appeal of human-evolution studies, again putting Creationism and anti-evolutionism on the back foot.

Thereafter the fortunes of Creationism matched those of US fundamentalism in general. While somewhat faltering in the later 1980s, following the televangelist scandals, it was sustained by an underlying apocalyptic Cold War political climate and growing doubts about the sufficiency of science to solve all humanity's problems.[17] With the end of the Cold War and the ensuing indefinite postponement of nuclear Armageddon, a more 'scientific' strand of ID theorizing finally began to make headway in the 1990s. This may be taken as initiated by Phillip E. Johnson's *Darwin on Trial* (1991) and the foundation in the previous year of the Seattle-based Discovery Institute by Bruce Chapman, George Gilder, and Stephen C. Meyer as a 'non-profit educational foundation'. This was followed by the formation of a pressure group, calling itself 'The Wedge', at a 1992 conference held at Southern Methodist University, which included Johnson, Meyer, Michael Behe, and William Dembski. The scientific ambitions of the movement were then clearly signalled in Michael Behe's *Darwin's Black Box, The Biochemical Challenge to Evolution* (1996), in which he claimed that the flagella of some species of bacteria were 'irreducibly complex' and could not have evolved (see below). This was cast as the 'irrefutable' example of design that was required in order to prove the necessity of design for science itself, and it was further bolstered in 1998 by William Dembski's *Mere Creation: Faith, Science and Intelligent Design*, the first of his on-going series of pro-ID books.

William Dembski, Professor of Science and Theology at Southern Seminary, Louisville, Kentucky, soon became one of the most influential ID theorists (although Meyer is intellectually the more impressive of the two). The most technically ambitious of his subsequent books was *No Free Lunch: Why Specified Complexity Cannot be Purchased Without Intelligence* (2001). A powerful 2002 on-line refutation of Dembski's argument by Richard Wein, and subsequent exchanges between the two, are downloadable from www.talkorigins.org/design/faqs/nfl. Dembski's argument is dauntingly technical for the uninitiated, but alongside innumerable demonstrations of the invalidity of mathematical and logical details, Wein claims to expose the entire work as a covert reworking of the traditional Argument from Design and 'God of the Gaps' arguments. He concludes: '*No Free Lunch* is completely worthless, except as a work of pseudoscientific rhetoric aimed at a mathematically unsophisticated audience which may mistake its mathematical mumbo-jumbo for genuine erudition' (p. 49).

Michael Behe's earlier (1996) *Darwin's Black Box, The Biochemical Challenge to Evolution* was more modest, challenging the adequacy of evolution only at the biochemical level. In particular he argued that the DNA specification for a particular form of bacterial flagellum, which acts like an outboard motor, is 'irreducibly complex', consisting of three distinct parts – a motor, a shaft, and a propeller. We will return to his concept of 'irreducible complexity' later. His argument was impressively complex, as even his most ardent opponents conceded. What this boiled down to was that the flagellum could not function without the complete set of DNA instructions, hence any less complete evolutionary precursors would be non-functional and could not survive.[18] In fact, other bacteria with simpler flagella were soon identified, and his account of the logic of evolutionary theory was, as opponents quickly showed, misleading. Two other biological phenomena were also cited as impossible to explain in evolutionary terms: the Krebs cycle (which governs the metabolism of carbohydrates, fats, and proteins) and the blood-clotting 'cascade' (how blood clotting in response to injury is triggered). Both of these phenomena are also extremely complex. Perhaps, though, the real underlying difficulty with design positions regarding these and similar phenomena is that it is only relatively recently that scientists have acquired the resources and research techniques to explore them. Unsurprisingly, then, scientific understanding is still at an early stage, and much remains unexplained. The evolution of bacteria species is rendered more difficult, of course, because they leave no fossils. Once more, the pro-design strategy thus closely resembles the long-abandoned 'God of the Gaps' approach, but aimed at the level of microbiology and genetics, rather than palaeontology.

By the early years of the twenty-first century the purportedly scientific ID camp was mounting a major assault, which continues, on evolutionary thought in particular and on the broader scientific cosmology in general. While clearly, and for the most part confessedly, religious in motivation, the proponents are, as they see it, attempting to engage science on its own terms. While many academic proponents of ID and 'Creation Science' avoid religious arguments as far as possible in their expositions and rarely espouse literal biblical Creationism, the boundary between the two camps is extremely blurred. We should remind ourselves at this point that while Creationism and ID theories are currently being promoted almost exclusively by certain Christian constituencies, resurgent fundamentalist positions in both Islam and Judaism are similarly committed to belief in the literal truth of the *Genesis* account.

The British scene has so far remained very different from that across the Atlantic. Few people in the UK are aware of the intensity and scale of the controversy in the United States. But there are signs of a change. As mentioned in the Introduction, of most concern have been recent suggestions that Creationism and ID should be included in school science curricula as alternative theories which should be respectfully treated. This has largely been a consequence of governmental encouragement of state-sponsored 'faith schools' and officially expressed sympathy towards the presence of religious interests within the state education system.

Summary

Clearly, Creationism (including ID) rests on two main theses, one religious, the other philosophical. The religious case is that if the Bible is the revealed Word of God, then what it contains must be literally true, and true Christians must, as a matter of faith, believe that the world was created in six days, that it 'fell' shortly afterwards, and was radically reconstructed by Noah's Flood some time later. This is primarily a position espoused by religious fundamentalists. The philosophical thesis is at heart the 'Argument from Design'. This holds that beyond a certain level complexity cannot arise, except as a result of intelligent design by an external designer which, it somewhat problematically assumes, must be a single agent and, in the case of the universe as a whole, has, therefore, to be God. It is perhaps significant that the former thesis is almost entirely a Protestant position, while the latter can draw support from some Catholics also, Michael Behe being a case in point.[19] The first of these positions will be tackled more directly in Chapter 5, but for the non-fundamentalist it is the latter which appears more convincing and thus deserves our initial attention.

2 Design and complexity: Paley's mistake

The Argument from Design

Our focus in this chapter is solely on the Argument from Design, which, as we saw, has been revived in an updated form by present-day American proponents of ID such as William Dembski and Michael Behe. The heart of the original Paleyite argument was that the existence of a designer has to be inferred from a phenomenon's complexity. This argument is, I hope to show, fundamentally flawed. First, however, an important preliminary point has to be made if we are to see the issue in its proper historical perspective.

To be fair, the emergence of the Argument from Design in the thinking of John Ray and his successors during the course of the eighteenth century was in large part a perfectly understandable emotional and religious response to the majesty, harmony, and wonderful variety of the universe as apparently being disclosed by the newly emerging natural sciences, from astronomy and physics to natural history. God's creation of the universe being axiomatic for all Christians, the new discoveries could hardly be read as other than a confirmation of that axiom and as evidence of levels of Divine ingenuity and genius even greater than those previously known. A philosophical formulation of this inference (rigorous and logical) was more problematic, and, despite the efforts of William Paley and others, cracks soon began to appear in it. It is now, as we will see, clear that the Argument from Design is fallacious. The vital point to be emphasized here, however, is that this 'classic' version of the Argument from Design differed from contemporary uses in one very important respect: it sought to provide a convincing argument for the existence of God by actually *deploying* the findings of science accumulated since Newton: it was not anti-scientific, although it was obviously poised against the rising tide of materialist and rationalist thought which was affecting European culture as a whole, and the thinking of many proto-scientists in particular. In short, it was a product of a very specific set of historical circumstances which rendered it plausible and appealing to a wide section of the Christian population. Moreover, it represented a way of maintaining the authority of religion which did not rest either on institutional authority (as did Roman Catholicism) or on supposed evidence of the supernatural (either subjective, as in mysticism, or external, as in cases of 'miraculous' cures following intense prayer). The Paleyite position, then, did *not* imply that 'supernaturalistic' explanations had a place in science itself. It was the entire cosmic ensemble which had been designed from top to bottom. What science was doing was unravelling how it worked and, in doing so, displaying the spectacular cleverness of God.

By the 1830s Paley's cosmology was in serious trouble on three fronts. Scientifically, the discipline of Geology, which had made only faltering headway in the previous century, was rendering the biblical time-scale unsustainable and casting doubts on the entire scriptural cosmology, including the Flood. This did not merely present a challenge to the literal truth of *Genesis*, but potentially opened up a space for non-purposive explanations for natural phenomena. The force of the Argument from Design implicitly rested on acceptance that the universe was created all at once, and all the fine adjustments of means to ends so evident throughout the Animal and Vegetable Kingdoms (as they would have said) were in place from the start. But once an extended time-span was introduced, the possibility was opened up of complexity somehow increasing over time by 'natural' means.

The second front was Theology, as we saw in the previous chapter. As the new German critical approach to Bible scholarship gained momentum, many devout intellectuals found it harder to sustain acceptance of the Bible as literally true in every verse and chapter. The Argument from Design was also unsatisfying, because it conflated Primary and Secondary causes – that is, God was depicted both as the initiating Primary Cause of the universe *and* as responsible for its detailed manufacture in every respect – i.e. its Secondary Cause as well. Given the mounting scientific evidence, would it not be wiser to confine His role to that of Primary Cause? This indeed became a standard argument of pro-science Christians throughout the nineteenth century, especially in regard to evolution. For many, the issue was now the more subtle one of the place of mind in Nature, which among other things could lead to the espousal of a Theism in which God is conceived as the ultimate ground of all existence, pervading and sustaining the entire universe, but not overtly visible. The validity of the Theist case was propounded very effectively by the Victorian Unitarian theologian and philosopher James Martineau (1805–1900), who argued that logically one had to postulate the existence of some kind of God-like 'Mind' in Nature. This question does, I believe, remain in some respects open, as I will try to show in Chapter 7.[1] The source of authority for religious belief was also being reconsidered: Martineau, for example, argued that it was indeed ultimately a matter of subjective experience, that it was *psychological* in nature.

The third front was philosophical, and it had been opened up by David Hume in his *Dialogues Concerning Natural Religion* as early as 1779. From this direction it became ever clearer that the logical force of the Argument from Design was, to say the least, extremely weak. In a sense the present chapter is continuing this tradition.

As observed in the previous chapter, in contemporary Creationism/ID the way in which the Argument from Design is used is almost diametrically opposed to that of original Paleyite Natural Theology. It is being used not to prove, or reinforce belief in, the existence of God by invoking the apparently self-evident presence of design in immanent fashion throughout the scientifically revealed cosmos, *but as a route for attacking that very cosmology.* What seemed self-evident to Ray and Paley has become the very point at issue. *Precisely because the self-evident presence of design has evaporated, the quest is to find something, anything, in the natural world which can be unambiguously demonstrated as having been designed by an external agency.* It is, despite protests by ID proponents and 'Creation Scientists', anti-scientific. We will have much more to say about this, but it is, I believe, essential to emphasize it here, since contemporary ID advocates often depict themselves as heirs and restorers of an unjustly rejected Paleyite tradition. They are not.

We may now get down to business.

Criteria for identifying Design

The first key point to be made is that *design and complexity are unrelated.* We all recognize that such simple things as screws, teaspoons, pins, string, flowerpots, and sheets of paper are designed. And we do not claim that such complex phenomena as volcanoes, weather systems, frost patterns, and weather-worn cliff-faces are designed. If the term 'design' is to mean anything, it must be possible to contrast it with that which is not designed, so we need criteria for distinguishing between them. Complexity as such cannot do so, because it would omit hooks and toothpicks – but maybe there are levels of complexity which simply have to be ascribed to design rather than 'chance' ('irreducible complexity', as Behe calls it)? But mathematicians have now clearly demonstrated how phenomena of great complexity can arise from 'chance'. There are whole sub-fields of applied mathematics such as Chaos Theory and Complexity Theory which explore just this issue. Chaos Theory is now fairly well known. Complexity Theory arose in the 1980s–1990s; a popular, rather anecdotal, exposition of it is Roger Lewin's *Complexity. Life at the Edge of Chaos* (1993). Complexity Theory suggests that there are inherent laws or principles according to which order 'crystallizes' within dynamic interacting systems. These, it is claimed, supplement or complement classic 'natural selection' in the evolutionary process.

Conceptually, the crux of the matter is not complexity, but the idea that 'design' implies three things: a designer (or a group of collaborating designers), the object or phenomenon designed, and *the purpose* for which the designer created it. (The designer must of course be external to the object itself – hence the power of Paley's watch analogy. Objects cannot consciously design themselves.[2]) These have nothing to do with complexity. We need not always know who the designer was, or what the object's purpose was, to reliably infer that it was designed. The designer may be long forgotten (as in the case of Stonehenge), but if the object possesses features in common with, or incorporating, other artefacts – being made of cast iron and wood nailed together, for example – it is a fair bet that someone 'designed' it, even if we now have difficulty in figuring out its purpose. The latter often happens with prehistoric artefacts – again

Stonehenge is an obvious example. *The triad of designer, object, and purpose is what is crucial, not the complexity of the object itself.* In essence there are two valid inferences to be drawn on the basis of this triad:

- given an object and a purpose for the object, we can infer the existence of a designer or group of designers; and

- given a designer (or group of designers) and an object which the designer (or group of designers) has created, we can infer that it had a purpose (even if we cannot figure out what it was).

In the case of Stonehenge, even though we appear to lack knowledge of both designer and purpose, we know that it was 'designed' by humans, (a) because it fits into a matrix of wider archaeological knowledge, and (b) because there are several quite plausible 'purposes' that it may have served. There are also no known 'natural' forces which could have created it – although there are probably those who would like to believe that extra-terrestrials were involved! One's ability to draw inferences is, of course, dependent on one's state of knowledge. To someone ignorant of archaeology and astronomy, Stonehenge would not appear self-evidently designed by humans, or even designed at all. The 'Giant's Causeway' in County Antrim (Northern Ireland) is a reverse case: now known to be composed of about 40,000 basalt columns of volcanic origin, it once seemed 'obvious' that it was manufactured, so the legend of its being built by the giant Finn McCool arose. In the case of Paley's watch, had it been found by a contemporary forest-dwelling inhabitant of, say, New Guinea, never in contact with Europeans, his or her conclusion (if the watch was working) might well have been that it was some kind of animal or, indeed, that it had been created by a god.

In short, if there is no object, then there is nothing to talk about; and if an object has no apparent purpose, we cannot infer a designer (nor indeed can we evaluate how well or badly it is designed, even if it is). Being 'designed' is not an inherent property of objects or phenomena but refers rather to their mode of origin; it cannot be deduced from the character of the thing itself, unless the existence of either an external designer/ designers or a purpose is already known or can be reliably inferred.

At this point we should note that in the original Paleyite scheme of things the Universe and its contents *did* have a clear purpose. The very point of it all was to supply 'Mankind' with 'his' needs, whether physical, moral, emotional, aesthetic, or intellectual. And in doing so it called forth from the believer's heart intense feelings of awe, reverence, gratitude, and humility. But that comforting anthropocentric universe is gone – a disappearance which caused much heartache and grief to thinkers in the late nineteenth century.

Where does this leave the modern ID argument? In some difficulties, for what confronts us in the cases of life-forms and astronomical phenomena is a situation where the existence of the designer is the very point at issue, the designer's purpose in making them is quite unclear, and the objects themselves are apparently self-sufficient, bearing none of the hallmarks of external artifice as we understand it, seemingly being created by natural physical processes. With human artefacts we can know or surmise how they were manufactured or crafted by an external human agency: how the pot is thrown, the wood carved, the whisky distilled, and the picture painted. But nobody has ever reliably witnessed any agency other than natural processes in play to produce any of nature's phenomena. (The 'eolith' controversy, which raged in the 1920s, is an interesting case of the difficulty in identifying human agency or 'design'. Some rather enigmatic stones had been found at the English seaside town of Cromer and near Ipswich which the archaeologist J. Reid Moir, supported by the eminent Sir Ray Lankester and fellow archaeologist Benjamin Harrison, claimed were extremely early stone tools, called 'eoliths' from the Greek 'eos' for 'dawn'. These have long been rejected as human artefacts, but the episode remains an interesting case study of scientific controversy and its resolution.) This highlights another problem with the ID case: what would ID look like in action? Some kind of micro-level telekinesis in play as the bacterial genome was adjusted? Can ID theory avoid what is called 'supernaturalism'? It is vital to its purposes that it do so, because science excludes such explanations on principle.

A further minor point which is nonetheless of some relevance here is that many earlier pro-design and theistic thinkers were also rather unhappy about the very notion of 'supernatural' miracles, because such events meant that God had had to intervene arbitrarily in His creation, implying a design flaw inconsistent with His Absolute Wisdom. Creationists may try to get round this by arguing that we live in a fallen world – or presumably a fallen universe. This, though, would be trying to have it both ways: impressively complex phenomena 'proving' the Argument from Design, and obviously badly 'designed' ones, such as natural disasters, 'proving' that the world is fallen.[3] It is, however, probably unfair to accuse many contemporary Creationists of relying on such a patently false argument.

Complexity, design, and ignorance

One root of the difficulty, I think, is that we are mesmerized by complexity as such. (I am leaving aside here quite what we mean by 'complexity' – which is not in fact as straightforward as it seems.) We just have a feeling, not based on any real evidence, that 'chance' as we call it simply *cannot* produce phenomena of the complexity that we find in nature, be they galaxies or the human hand. Instead of considering that it may be our understanding of 'chance' that is faulty, and admitting our ignorance, we race through a series of reflex inferences. If it cannot be 'chance', it must be 'design' ... if it must be 'design', there must be a designer ... if there must be a designer, it has to be God. And once we have introduced the word 'God', we automatically equate this with 'God' as conceived in our particular religion. When someone points out that God's purpose in all this is utterly obscure, or that actually a particular piece of anatomy is rather badly 'designed' (for example, the fact that the blood vessels in the human eye, unlike those of the octopus, are in front of the light-receptors, or that we possess a useless appendix which does nothing other than render us susceptible to appendicitis), we respond that this very incomprehensibility is just more evidence of how superior God's wisdom is to ours. Another interesting example of bad design, by the way, is how our upright human posture renders us especially vulnerable to spinal problems.[4] As Krogman (1951) said more than sixty years ago, 'A long dreary catalogue of physical disorders – muscular strains, prolapses, hernias, backaches and disorders of the feet and legs – reflects part of the price we pay for walking upright' (p.56). It is a little difficult to reconcile this with the notion that God literally made us in His image.

Well, again you cannot have it both ways, invoking both seemingly brilliant *and* seemingly botched 'design features' as equally weighty evidence for divine wisdom. Such a mode of arguing simply assumes what it was supposed to prove in the first place, i.e. the existence of a designer. This is a logical fallacy known technically in the philosophical trade since ancient times as *petitio principi*, and more commonly as 'begging the question'.

Assuming the incomprehensible to be supernatural, rather than an index of one's ignorance, is a common human vice. I have to admit bafflement as to why people are so easily convinced by the 'paranormal' feats of stage magicians and illusionists: I am quite reconciled to the fact that if someone devotes years of training, study, and ingenuity to perfecting a skill then I will probably be unable to understand how they do it – especially if achieving such incomprehension was their goal in the first place! How pianists can play arpeggios, and indeed how my computer works, are mysteries to me, but I do not jump to the conclusion that something paranormal or supernatural is going on. Similarly with the feats of nature, I can happily concede that the energy transformations at the heart of all things are beyond my comprehension, although perhaps no longer entirely beyond that of mathematical physicists.

In sum, the appeal of the ID argument rests less on its logic than on our perennial inability to think about and clearly analyse the ideas and concepts we are using, in this case the very concept of 'design' itself. We invalidly confuse 'design' and 'complexity' and are reluctant to admit that the natural world exhibits none of the features which characterize 'design' as usually understood. (The term 'chance' could also do with some re-examination, in fact.)

The real nature of design

Yet there is another major flaw in the argument, less abstract, and hence in some ways even more telling. In reality, complexity frequently signifies the *absence* of a single overall designer. *The choice then, when faced with complexity, is not between chance and a single conscious designer, but between chance, a single designer, and a multiplicity of designers who may or may not have consciously intended the actual total outcome.* Ironically, Paley himself, in a passing throw-away clause, concedes the possibility of a multiplicity of 'artificers', but fails to see how this potentially subverts his entire argument.[5] And a correct application of the analogy with human design would surely suggest that the last of these is more likely, for complexity in human-designed phenomena usually arises when numerous agencies with different individual purposes interact in such a way that all their variously competing and complementary ends can be met. The upshot, as in the global economic system, is a dynamic and highly complicated 'product' characterized by a multiplicity of *ad hoc* arrangements – in this case, legal 'checks and balances', regulations, international trade agreements, committees, and 'regulators', with different kinds of power and responsibility which they may or may not be able to exercise effectively, and so on – but no single 'designer' is in control, and no single 'purpose' guides it. The irrefutably complex global economic system has been 'evolving' for at least four hundred years, certainly since the coffee houses of Restoration London in the late 1600s. As far as human-made objects and systems are concerned, complexity thus tends to indicate the *lack* of a single 'designer'! Paley himself also seems to have conflated the roles of 'designer' and 'maker' under the term of 'artificer', somewhat surprisingly for someone writing in the wake of Adam Smith's elaborate treatment of the 'division of labour' in his economics classic *The Wealth of Nations* (1776).

Evolutionists believe that this is also the case in the natural world, exemplified by complex ecological systems in which multitudes of life-forms, each with different survival requirements, have to find strategies of co-existence. As they do so, they in turn change the situation in such a way that other life-forms have to make new adjustments. And sometimes they fail and become extinct, just as companies sometimes go bankrupt. The sheer complexity of ecological systems can surely most convincingly be understood, like the current global economic system, as the unintended compromise outcome of a multitude of individual agencies striving to meet their own ends. This is far more plausible than viewing them as the one-off achievement of a single transcendental 'designer' who, in a unique fit of creativity, carefully crafted everything from the life-cycle of the liver-fluke to the flightlessness of the ostrich. (Being charitable, one should perhaps bear in mind that Paley knew nothing of germs, viruses, bacteria, and microscopic parasites, so he was able to view organic life-forms in the idealized fashion required for a universe divinely adapted to human needs.)

The internal complexity of life forms can be similarly accounted for in this way. Physiologically, organisms have to operate in a way which reconciles a wide range of needs: nutrition, reproduction, ability to cope with immediate environmental conditions, and so on. Their morphologies represent strategies for simultaneously meeting all these needs in a viable fashion. From the evolutionary perspective, these strategies gradually accumulate (and sometimes disappear again) as survival conditions change. Some features are, as it were, 'knock-on' effects of others. Sometimes adaptations have accompanying features which fortuitously provide a basis for a subsequent adaptation to serve purposes for which they were not initially 'designed', as S. J. Gould and R. C. Lewontin pointed out in their famous paper 'The Spandrels of San Marco and the Panglossian Paradigm: a Critique of the Adaptationist Programme' (1979). Evolution is very adept at 'recycling'. None of this implies some high-level 'intentionality' on the part of the organisms. Many species have entered blind alleys and become over-specialized, or rendered themselves very vulnerable to slight environmental shifts. The Giant Panda is a good example, being reliant, as it is, on large quantities of nutritionally low-grade bamboo. Such species can easily become extinct. Others, like rats and humans, have become highly 'generalized'.

So, in the light of this, what of Paley and his famous watch? It was *not* after all the creation of a single designer, but the 'evolved' product of several centuries of creative effort by generations of clockmakers and watchmakers, striving to find ways of reconciling the competing demands of factors such as portability, accuracy, expense, robustness, and aesthetic appeal to customers. Some of its features (like having a numbered dial and escapement mechanism) were not designed by its maker at all, but were 'inherited' from his or her professional ancestors; while others, like the finish on the watch-case, would probably have been supplied by someone else, who may or may not have personally designed them. The individual watchmaker's own creative contribution to its design (if any) may have been confined to one subtle innovation in its internal mechanism to improve its time-keeping accuracy.

Virtually all human designing is a social process in some important respect. Not to pull any punches here, *none* of the designed human artefacts that we now routinely use is the sole creation of a single 'creator' – even craft potters rarely design and build their own kilns from scratch (and pots as such were invented aeons ago). Computers, cars, aircraft, books, furniture ... virtually everything that we make and use has a longer or shorter history of 'evolving' design in which hundreds, sometimes thousands, of individuals participated somewhere along the line. And in every case their efforts were responses to the needs and requirements not only of themselves but of others too. In the majority of cases the core 'purpose' of their actual makers, who may or may not have had a hand in their design, is really to earn a living. And it should be remarked that human designers generally strive as far as possible for simplicity, not complexity (except when showing off their expertise). All of which serves to weaken still further any analogy between human artifice and divine design, since God was not presumably involved in any social relations when He made the universe, nor did He need to earn a living. If the design analogy does have any theological implications, they would appear to be in favour of polytheism rather than monotheism.

It is, incidentally, perhaps no coincidence that Paley's Argument from Design emerged at the height of romantic exaltation of the figure of the lone heroic artistic genius. But even Leonardo da Vinci and Michelangelo needed sponsoring popes and princes – as well as assistants on the ground. The 'old masters' often also delegated some of the more boring bits of a picture to an apprentice in their studio, who perhaps had shown a flair for painting grass. More immediately one might note that 'design' was itself a very trendy topic in late eighteenth-century England, the Royal Society of Arts having been founded in 1754 explicitly to promote design in the arts and manufacturing.

Paley's error: a summary

Taken together, chance (in the case of inorganic phenomena) and chance plus unconscious multi-agency 'design' (in the case of organic ones) thus appear to be quite sufficient to account for the physical universe. To sum up:

- 'Design' implies three things: (a) a designer or designers, (b) a designed object, and (c) the designer's purpose for that object. In the case of natural phenomena, all that we know exists are the objects and phenomena themselves. Since they do not discernibly serve any external designer's purposes, we have no reason to assume that such a designer exists.

- Complexity and design are unrelated. There are innumerable very simple designed human artefacts, and many very complicated undesigned phenomena. The more complex a human artefact is, the *less* likely it is to have a single designer, usually being the product of the labours of many 'designers', and indeed non-designers, in the way outlined above. This is essentially a social process. Complex natural phenomena of an organic kind are similarly usually the outcome of mutual, quasi-social, interactions between many autonomous life-forms or responses to specific environmental contingencies. This need not imply any conscious knowledge, intentionality, or foresight beyond the contingencies of immediate survival needs – and even then some get it wrong.

- As far as 'chance' is concerned, complex phenomena can also easily arise at the intersection of different forces such as sea, rock, wind, glaciers, electrical fields, temperature variations, and the like. Life itself was most likely one of them.

There are thus, on the face of it, no grounds at all for postulating the existence of a single supreme 'designer' on the basis of the complexity of natural phenomena. Indeed, despite the initial plausibility of Paley's example of the watchmaker, this is quite illogical: first in terms of the necessary criteria for using the concept 'design', and secondly because the analogy with human design simply does not hold up. So what was Paley's mistake? Fundamentally that he failed to ponder properly on both the meaning of the term 'design' and the real nature of the watch-making industry.

Some further complications

But is there really no sense at all in which complexity and design are related? Unfortunately for Creationism, there is just one: the capacity to engage in intelligent designing is dependent on the prior complexity of the designer. In other words, complexity *precedes* the capacity to design. In the only major case known to us – ourselves – human designing has emerged from a blend of social, biological, and psychological complexity. When something resembling it occurs in other species, it too is only possible because they have achieved some form of 'complexity': social complexity in the cases of ants and bees, and in higher primates socio-biological complexity verging on our own. The problem for Creationists is that, pursuing this line of thought, God must have already achieved the requisite level of complexity *before* becoming capable of engaging in intelligent designing. If, as is claimed, complexity is itself created by intelligence, then which intelligence created God's complexity? This obviously leads to an absurd infinite regress. The key point is simply that complexity precedes design and is not its necessary consequence (although it can be).

A further relevant issue should be mentioned before leaving this topic. The Christian attitude to Nature has long been self-contradictory. On the one hand (possibly as a consequence of the Fall), the Earthly creation has been seen as ageing and decayed, a realm inferior to the spiritual domain. The best that can be said is that God created it to serve human needs until the Day of Judgement, giving Adam, Eve, and their descendants complete rule over it. This is consistent with Strong Creationism and Pre-Millennialism. It is a position which has led many to blame Christianity for the exploitative domination of Nature by Western cultures which has caused the current environmental crisis. By contrast, many Christians now espouse a 'stewardship' position on environmental issues, St. Francis of Assisi figuring prominently as a role model. This was also, however, surely facilitated by the eighteenth-century Argument from Design and the emotional grounds for its appeal. Although often given a thoroughly anthropocentric spin – as in the Bridgewater Treatises – it could easily echo Romantic Nature worship. I would therefore like to pose the question: how important actually is the Argument for Design for Pre-Millennialist Strong Creationists? The core of the Strong Creationist case would seem to be the *Genesis* story as literally true. The beauty of Nature does not come into it and is even an impious distraction. At the beginning of this chapter I described how different the aims of the eighteenth-century and early nineteenth-century exponents of the Argument from Design were from those of its present-day advocates. From the early nineteenth century onwards, a clear split emerged in the Christian camp. On the one side were those who sought compromise and reconciliation between science and religion, who rejected the original Argument from Design in favour of one in which God was a 'Primary Cause' only, and who abandoned the extreme anthropocentrism of the Bridgewater Treatises. They might dispute specific issues with the scientists, but they saw warfare with them as profoundly mistaken. On the other side were those who really had little interest in the Argument from Design but rested their case on a simple fundamentalist reading of the Bible as an authority trumping all others, anticipating today's Strong Creationism. This may be seen as a desperate fall-back position in the wake of the failure of the Argument from Design. Science was now the enemy.

I cannot see how these two camps are reconcilable. On the face of it, it might look as if the modern ID camp, the 'Creation Science' school, are heirs of the former tradition, wanting to 'reconcile' science and religion, but (as I pointed out before) this is not the case. They are better seen as in a new position which is emotionally close to Strong Creationism, but – understanding that this cannot be defended in the current climate – they have fallen back on the hope of vindicating the obsolete Argument from Design, even while misunderstanding its original character and thrust as proving the existence of God, not explaining the world. This is perhaps a subtle distinction, but it goes to the heart of the whole issue. For those espousing the Argument from Design, scientific laws and scientific explanations of how natural phenomena worked were the evidential base for God's existence. They were not, like Behe, concerned with identifying specific instances or examples of purposive design to explain phenomena otherwise scientifically inexplicable. For them the entire cosmic show was purposefully designed. *The Argument from Design was a way of looking at the world, not a way of researching it.* Ironically it is now the Strong Creationists who come closest to this position, but on the basis of blind faith, not because of Paleyite argument. If this analysis is valid, the Creationist/ID position is full of far more fundamental internal conflicts than is usually acknowledged by its exponents or detected by its opponents.[6]

There are, however, somewhat more profound issues involved than the argument that 'complexity signifies a single designer'. So far we have only lanced one central conceptual boil, rather than providing a fully sufficient refutation. In particular we have not addressed the 'cosmogenic' question: where did the universe come from in the first place? It may be conceded, for example, that human design always takes the form of manipulating pre-existing matter, but surely we need to invoke something analogous to design to account for the very existence of matter (or, more strictly speaking, energy) itself and the complex, delicate laws which govern it? It is with this conundrum that the next chapter is concerned.

3 Rules, laws, causes, and explanations

The 'Laws of Nature' are undoubtedly extraordinarily subtle. If certain mathematical constants characterizing the behaviour of the basic forms of energy that hold atoms together and stars apart were even marginally different, the entire cosmos would, we are reliably informed, either collapse or dissipate. Equally remarkable are the laws of chemistry and biology that enable terrestrial eco-systems to maintain their balance, or animate organisms their integrity and survival. Unbelieving scientists are no less awestruck by all this than are the religiously devout. Two questions are thus raised: the general one of 'Why is there something rather than nothing?', and the more specific question of 'How did these laws originate?'.

There are two concepts in play here which are of central importance to the topic of this book: 'law' and 'cause'. Before tackling them, a few preliminary remarks are in order.

Even if it transpires that we did feel compelled, at the end of the day, to grant that some agency was ultimately responsible for creating the universe and the laws governing it, this would not require us to call it 'God' in any senses of the term that we are familiar with in traditional monotheistic religious systems. And in using the term 'design' to describe this, we must remember that this is but a crude and misleading analogical use for want of anything better. Creation from absolutely nothing is a quite different order of achievement or event from creating from existing material. Dembski introduces the notion of an 'Unembodied Designer', but this is really quite incoherent, and he falls back on that familiar favourite of modern speculative pseudo-scientists, 'the quantum level', as the site where this agency operates. In any case, this originating agency apparently disappeared entirely from the external universe as soon as it began. Once running, the cosmos was seemingly left entirely to its own devices. Nothing physical happens in it which cannot be accounted for by these 'natural laws', at any rate in principle.

The God concept is an inexhaustible topic of debate which cannot be adequately dealt with here, but it should be stressed that nobody has ever produced a really coherent account of what the word 'God' refers to. In the present instance, the creating agency must be so utterly remote from us in time and character that its existence (if we can even use that word for something which, since it produced *existence*, cannot presumably itself physically exist in the usual sense) can have no conceivable bearing on human affairs. This is not a novel point. Theologians and religious believers have always oscillated between two poles in their efforts to characterize God, as will be appreciated by readers of Karen Armstrong's extremely valuable *A History of God From Abraham to the Present: The 4000-year Quest for God* (1993). On the one hand, It (or He or She) becomes so transcendently absolute, wise, and incomprehensible that God turns into a meaningless abstraction of no clear relevance to humanity; on the other hand, He (usually) is a transparently obvious projection of local cultural concepts of absolute power and rulership. All the monotheistic religions have striven to find a middle course between these extremes, but they have never succeeded in reaching a consensus on where it lies. This does not imply that the concept 'God' is meaningless. From a Jungian point of view, for example, its eternal and inexhaustible indefinability and mysteriousness is the essential point of the 'God-archetype', as of all such symbols and concepts. In short, it would lose its whole point if it *could* be neatly and rationally defined.

But do we in fact need to invoke some such transcendental purposive agency to account for the cosmos and its laws?

Laws

In calling the regularities of nature 'laws', we are of course exploiting the analogy with human laws, consciously designed to regulate our behaviour and conduct – be it how civil society is organized or the way in which a game is played. But there is an obvious difference. Human laws can be broken; natural ones cannot. Human laws can also be changed, but again natural ones are, as far as we can tell, eternal and universal.

Some Strong Creationists have rather bizarrely tried to challenge this, proposing tortuous alternative interpretations of the basic laws of physics and astronomical evidence to save the six-day creation, 'Young Earth' hypothesis. These efforts can hardly be taken seriously, but readers may wish to check out Mark Isaak's 'Index to Creationist Claims' on the talkorigins.org website referred to in Chapter 1. To return to the point, the built-in connotation of deliberate conscious formulation which the ordinary sense of the word 'law' possesses can be carried over, almost unnoticed, into its analogical use.

To grasp this more fully it is also essential to understand that natural laws are not *causes*. Human-made laws might on occasion cause people to behave in certain ways (including driving on the right-hand side of the road) and break them in defiance, but a scientific law is simply a neat statement, typically a mathematical equation, which *describes* an empirically observed regularity. Scientific laws do not exist as external, separate, agencies controlling the phenomena that they describe. They are not like some covert cosmic police force compelling physical phenomena to exhibit certain properties. There is no 'law of gravity' independent of gravity. One might summarize this point by saying that whereas scientific 'laws' are empirical propositions – and thus true or false – human laws cannot be true or false, only adhered to or broken.

In mathematics things become very strange indeed, since there are no numbers in nature. Nobody ever discovered the number 3,872 out there in the woods, or the square root of 27 under a stone. Numbers were invented by humans as a tool for assessing quantity and, among many other things, for monitoring more precisely the regularities in external nature – how many days there are in a year or between full Moons, for example. The sense in which numbers may be said to 'exist' remains a matter of profound philosophical debate into which the author is unqualified to enter, and it is in any case peripheral to present concerns. All I wish to do here is to indicate, albeit crudely, the nature of the problem and its bearings on the issue at hand. Despite being a human invention, numbers and abstract geometrical forms are law-governed, quite independently of human wishes and intentions. In Euclidean space the angles of a triangle always add up to 180° (or the equivalent if a different degree-numbering system is used). This is true whether any triangles are present or not, *and it does not happen over time*. Nothing *causes* them to have this property. And, whether you like it or not (although it is hard to imagine why you wouldn't), 2 x 2 = 4. Another, related, example of a human invention governed by apparently objective rules is formal logic. No doubt about it, 'If A is bigger than B, and B is bigger than C, then A must be bigger than C'. Illustrious and ingenious mathematicians and logicians have taken their arts into realms of abstraction which are a mystery to the rest of us, encountering and overcoming many paradoxes, ambiguities, and alternatives to existing approaches in doing so. It is still possible to make mathematical 'discoveries'. But 2 + 2 equalling 4 and A's being larger than C, in the examples just given, remain unassailable 'laws'. I concede Paul Feyerabend's argument in his provocative *Against Method* (1975) that there can be historical shifts in what are considered to be logically sound arguments – the scientific rejection of the validity of the Argument from Design in the nineteenth century is actually a case in point – but this cannot, I think, apply to the basic laws of formal logic. These shifts typically arise from changed scientific understanding. This can affect both the empirical validity of the premises on which a logical argument has been based and also the acceptability of specific concepts employed. The formal validity of the principles of logic nevertheless remains unaffected.

What this brief meditation suggests is the possibility that natural laws are more akin to the laws of logic than to the laws of banking. Perhaps they describe the way things *have* to be, the only possible way they *can* be. They certainly describe the way things *are* rather than, as human laws do, how they *ought* to be. To infer from this that some external agency must have *created* them thus becomes a rather dubious logical move. Humans invented mathematics and logic, but they did not invent their laws; these, they discovered, possessed an independent, 'objective', 'natural' character of their own. But it is surely rather odd to propose that God somehow arbitrarily decided that the universe would best be served if 2 + 2 = 4, or rejected the possibility of A being smaller than C even though it was bigger than B and B was bigger than C, as if there were other options. Logical and mathematical impossibilities and contradictions are not the result of God in His wisdom deciding, after due consideration, to rule them 'illegal'. And, one should stress, it is precisely by deploying the 'laws' of mathematics and logic in relation to natural phenomena, by measuring and modelling them, that the natural laws of the physical universe were discovered.

How though does this bear on ID in relation to the *origins* of natural laws? One difficulty that it raises is for the very notion of 'origins' itself, for natural laws exist only as human-made descriptions of observed regularities and cannot in themselves have a controlling relationship as independent agencies or forces over the phenomena that they characterize. Moreover, as just noted, when we turn to the laws of mathematics and logic, the very instruments by means of which it became possible to formulate such regularities, we find that (a) they are quite atemporal, or somehow outside time, and (b) they have a quality of absolute necessity about them which renders the idea of their being cleverly created rather peculiar.

This does not get us very far with the 'cosmogenic' or origins issue as such – why there is something rather than nothing – but does seriously weaken the notion that 'natural laws' are themselves evidence of a conscious intelligent designer. Understanding why $2 + 2$ *must* equal 4 requires a modicum of human intelligence. The *fact* that $2 + 2 = 4$ does not. It strikes me that, since we do have to draw on analogies, it would be preferable to draw them from the laws of logic or mathematics than from the laws of the land, and the natural law $E = mc^2$ patently bears a closer resemblance to $2 + 2 = 4$ than it does to 'Thou shalt not steal' or 'Keep off the grass'. And natural laws cannot, as just explained, in themselves be temporal causes, although we naturally often talk is if they were. It is not the *law* of gravity that causes apples to drop and planets to stay in orbit: it is gravity itself – the 'laws' of which Newton subsequently discovered and mathematically expressed (and which his successors have subsequently refined). Mathematical natural laws and constants, unlike purely formal mathematical ones, were indeed empirically discovered, but this does not alter the fact that they have more affinity with these than with human-made laws and regulations (which are never *discovered*, only made). After all, as previously noted, it was through the very media of mathematics and logic that natural laws were generated. We have then to turn elsewhere, to the concept of 'cause', in order to address the question of origins.

Causes

The following section is necessarily a condensation of some extremely complex issues. The nature of causality, with the associated concepts of 'force' and 'agency', has been one of the most arduously debated philosophical issues of the last two and a half centuries or more. Into this I will not venture. Even more than 2,300 years ago Aristotle felt it necessary to identify four types of cause, in a work now known as *The Metaphysics* (so called because it was appended to another work, *The Physics*). These causes are referred to as Material, Efficient, Formal, and Final. I will have occasion to return to them in the next chapter, but I do need to say something here about Final Causes.

The term 'final cause' refers to the reason why something was made: what it was made *for*. Such explanations are technically called 'teleological' (i.e. 'purposive') explanations. This category of cause proved the most difficult to apply to the natural world. Organic life in particular apparently contains innumerable internal components, such as roots, lungs, livers, and hearts, which clearly serve a purpose, as well as displaying purposeful behaviours like blossoming, nest building, mating displays, and hibernation. And because a phenomenon serves a function in the maintenance of an organism's life and the survival of its species, initially it would seem logical to conclude that this function was its 'final cause', the work of a purposive agent of some kind. But for science this yielded a paradox, seeming to imply that something in the future is causing something in the past at the very outset of the chain of 'efficient' causes (or 'cause–effect' sequences) before the thing itself exists and becomes able to perform these functions. And causation from future to past runs contrary to the uni-directionality of time, on which, until certain very recent arcane developments in particle physics, everyone agreed. An effect cannot precede its cause. So the pressure in science, following the near-universal scientific rejection of the Argument from Design around the 1830s and 1840s, was always to find ways to eliminate the need for final causes and seek alternative kinds of explanation.

It was quite successful in this, most notably in evolutionary theory after the 1850s, and in introducing the concept of 'negative feedback' in the mid-twentieth century. The latter underpins all the basic principles (such as programming) in computing, robotics, and other apparently 'purposeful' artefacts. 'Negative feedback' is quite un-mysterious, occurring in the natural world as well as the manufactured world: for example, the 'homeostatic' regulation of bodily functions such as temperature. At its simplest it is what happens whenever a phenomenon has effects which then inhibit that phenomenon. An example in the natural world would be when growth in population of a prey species leads to an expansion in the predator population which then reduces its population again, resulting in a long-term balance between the two. This does not mean that human activities at least are not guided by 'purposes' deliberately aimed at achieving future states of affairs. Such a claim would be absurd. It does, however, mean that psychologists have been able to treat this in a way which avoids reverse causality. Final causes in this sense are presently existing plans and projects which we formulate 'in our own minds' and strive to implement. The outcome is unknown until it happens. This kind of account, its advocates claim, nevertheless eliminates the need for teleological explanations – which are anathema to orthodox science – in the traditional sense. Even so, the feeling that somehow Final Cause explanations are needed for everything in order to provide a complete account in Aristotelian terms dies very hard. The serious philosophical issue that this raises is, however, about the status of the concept of 'agency' and whether it is a unique property of humans. Again, this is too complex and technical to address here, although we will be returning to the point elsewhere.

A rather different type of Final Cause explanation has arisen in theoretical physics over recent decades: the 'Anthropic Principle'. There are several versions of this, but they all hinge on the notion that the universe necessarily took the form that it did in order to be comprehensible to humans, or rather to human-type intelligence. I shall return to this in Chapter 8. What is relevant to note here is that the terms in which the Anthropic Principle is being articulated and theorized in no way draw on religious, let alone specifically Christian, ideas or arguments. It is of course highly controversial, but it has arisen within orthodox science and is pursued with scant reference to the cases proposed by ID exponents and Creationists.

Aristotle's classification remains very handy, but his categories are rather more problematic than he appreciated. Everyday uses of the word 'cause', especially when dealing with natural phenomena, usually refer to his 'efficient' causes: the cause–effect sequences which brought something about. Scientific knowledge largely enabled us, in principle at any rate, to assimilate 'material causes' (the 'matter' of which things are made) into this category. One consequence of this was the rise of 'determinism': the idea that everything is entirely the result of blind cause–effect sequences and can be no way other than it is.

This is an over-simplistic way of describing the historical course of the argument, however. What the elimination of Final Causes had also done, as just indicated, was to subvert the need to postulate an *agency* actively at work behind the causal chains. For many scientists, following on from the argument of David Hume, the term 'cause' really only referred to regularity of succession. There were problems with this, to be sure, but they did not prove conceptually insuperable for scientists and materialist philosophers of the early nineteenth century. The upshot of all this has, of course, been the famous, or notorious, Free Will versus Determinism argument, which continues to this day. One reason why final (and, as we shall see later, 'formal') causes have remained popular is that they apparently offer a way round this. But our image of cause–effect determination remains rather crude. We tend to think of single chains of cause and effect, isolated unilinear processes like snooker balls bumping one into the other. Actually things happen in real time in the context of the totality of such sequences, occurring at the intersection of innumerable 'cause–effect' chains (even snooker matches). There is a strong alternative philosophical tradition which has always challenged the simple 'chain' image in favour of more holistic approaches (for example, A.N. Whitehead's 'process philosophy'[1]). It is at this point that we need to ponder on another concept: 'explanation'.

Explanation

There is a danger here of wandering off the main track of the argument, but a few observations are necessary at this point (see G. Richards,

2005, for a fuller discussion[2]). In particular, even many highly astute philosophers have tended to overlook one obvious psychological fact: explanations are produced in response to puzzles. And, while the person requiring the explanation may have all kinds of criteria for finding an explanation acceptable, in the final analysis the 'good' explanation is one which leaves the person requiring it no longer feeling puzzled. There are two minor exceptions to this generalization, but they do not seriously affect its broad validity. These are (a) in, for example, teaching situations when a student is asked for an explanation in order for the teacher to assess the student's comprehension of the matter in question, where the puzzle is 'Does the student know the explanation?', not the matter to be explained itself; and (b) in situations where the question is really a rhetorical expression of emotion ('Why did he die?', 'What have I done to deserve this happiness?'). What this suggests is that we need to analyse how things become puzzling. When we do so (a lengthy business, into which I will not now venture), we quickly see that a single object, phenomenon, or event can be puzzling in a vast variety of ways.

Aristotle's four types of cause can be viewed as representing one (rather over-tidy) way of categorizing types of puzzle, but his schema cannot really tackle their sheer multiplicity. In short, we can usually identify a whole host of cause–effect sequences of quite varied kinds which potentially explain something. *And the one on which you concentrate is that which resolves your particular puzzle.* Consider the question 'Why is that tree there?' Is this being asked because the tree is of a species not usually found in this region? Or because all the other trees have been chopped down? Is it because you want to know if it was planted deliberately (if so, why?) or grew naturally? Or perhaps you want an explanation in terms of soil conditions, or how its seed was buried by a squirrel which then forgot it. Why that specific tree is in that specific location at this specific point in time cannot therefore be answered in its totality in terms of a single cause–effect sequence. There is no single 'Explanation', privileged and basic to all the rest. You may need to range across economic, horticultural, social, and meteorological history, as well as botany, to get near to a comprehensive account. But in reality we ignore nearly all of these, because we are puzzled about it from only one direction. 'It was planted by the Third Earl to commemorate his favourite horse' might be all you want to know.

What has all this to do with ID and Creationism?

Puzzles

The answer is that it draws our attention to the need to ask an absolutely crucial question: *What puzzle, or puzzles, are ID and Creationism intended to remove?* Now not all puzzles are equally valid in the terms in which they are posed. Sometimes they are based on a misunderstanding, and the way to remove them is to tackle this. They may be based on a simple misperception: 'Why is that man staring at me?' – 'He's not, he's looking at the suit in the shop-window behind you' (or even 'He's not staring at anything actually, he's blind'). In fact, a very large class of puzzles can be removed by explaining how they originate in an error in the puzzled person's assumptions, pointing out how he or she has somehow categorized the phenomenon wrongly. There is, however, a particular kind of puzzle to which we are all prone, and perhaps we would not be human were we not.

Until the middle of the previous century, many anthropologists and psychologists (though not all) claimed that pre-literate and tribal peoples typically exhibited what these Western experts considered to be illogical or pre-logical thought, often termed 'magical thinking'. The most elaborated version of this was that propounded by C. Lévy-Bruhl, on which he backtracked somewhat towards the end of his life. His most influential exposition was *Primitive Mentality* (1923, first published in French in 1922). A typical example would concern a 'primitive' man being hit on the head by a falling tree branch. While conceding that the branch was rotten and that a gust of wind broke it just as he passed beneath, he would nevertheless blame the event on witchcraft or some malevolent spirit. This struck 'civilized' observers as ridiculous. The rottenness of the branch and the presence of the wind were all that were required to explain the event. Ironically they failed to realize that we all sometimes think like this, not least if we are religious. We still seem to feel unsatisfied by the obvious 'efficient cause' of a misfortune (or piece of luck). We cannot help asking 'That's all very well, but why ME, and why NOW?' And although not invoking witches, we are quick to seek an explanation in terms of reward or punishment originating from some divine or demonic source who has us in its sights. This, as demonstrated by numerous public responses to the events of 9.11 and Hurricane Katrina by US preachers such as Pat Robertson, is still of course the invariable response of fundamentalist and evangelical Christians, as well as believers in other creeds, to any unforeseen natural or civil disaster. A curious example came to my attention soon after I started writing the present book. In relation to the outbreak of the Israeli–Hizbollah conflict in July 2006, Rabbi Lazer Brody, author and Dean of the Breslov Rabbinical College in Ashdod, Israel, blamed the conflict on a planned gay and lesbian parade for World Pride Day in Jerusalem, saying that it 'soils the camp of Israel with impurity, and pushes away the divine presence and protection', and claiming that 'When God's presence is in the camp, nothing can happen to the Jewish people. But if the Jewish people bring impurity into the camp of Israel, this chases away God's presence.' He was supporting another rabbi, Pinchas Winston, described as 'a noted author, rabbi and lecturer based in Jerusalem': 'Why does this war break out this week, all of a sudden with little warning? Because this is the exact week the Jewish people are trying to decide whether the gay pride parade should take place in Jerusalem or Tel Aviv'.[3] And a year later an Anglican Bishop responded similarly to severe flooding in the sin-ridden and decadent English county of Gloucestershire (*The Daily Telegraph*, 7 July 2007).

This tendency can be reinforced by what C.G. Jung called 'synchronous' events, more commonly known as 'meaningful coincidences'. Somehow or other, two or more of the multitude of cause–effect chains at work in the world (chains which we have no reason to believe are connected) suddenly intersect in our experience in a way which we cannot help but feel to be meaningful, as if something out there was trying to get a message through to us. The more hard-headed materialist scientist or person who has never had such an experience may try to dismiss it summarily as 'just coincidence', but this is in many instances quite unsatisfying (and in fact is another example of 'begging the question').

It is, admittedly, usually hard to know what to do with these events: they do not seem to demand any action and they often relate to quite minor matters. Occasionally, however, they can be quite overwhelming in their felt profundity, sometimes even resulting in religious conversion. Fairly often they are oddly helpful, as when you chance upon a book which is exactly what is needed for your research, although you have never heard of it before, and the next day you discover an article in a newspaper about the author.[4] Even so, it remains possible that we currently lack sufficient grasp of how all cause–effect chains inter-relate in real time to be able to identify the source of such 'coincidences'. It has also been rightly pointed out to me that a world without coincidences would be even weirder! (Geoff Bunn, pers. comm.)

I believe that the puzzle that ID and Creationism are attempting to resolve is at heart a puzzle of this order. It is a projection on to the cosmos as a whole of the kind of bafflement that most of us can experience when something very bad or very good or simply odd, like meaningful coincidences, happens to us. Instead of asking 'Why me? Why now?', we find ourselves wondering 'Why the Earth? Why *us*? Why anything at all?'. And the only kinds of satisfying explanation that we can come up with are those that postulate the existence of a transcendent agency (which we label God) in whose purposes, albeit utterly incomprehensible, we in some way centrally figure. The universe *must* have a Final Cause, and human affairs *must* have a meaning within that. Floods and earthquakes are punishments like the branch falling on the 'primitive' man's head, and if our society flourishes, that is our collective reward. We want to believe this, despite the lack of any visible differential in the levels of sin and piety between those who are involved in such events and those who are not. At heart it is about giving the world, and our lives in it, *meaning*.

The problem is that this way of experiencing the universe is not shared by everyone, and many, some of whom at least understand its appeal, quite happily do without it at no cost to their moral standards or enjoyment of life. They can accept that sometimes things happen which they cannot explain – but then they never assumed that they had the answer to everything. It is quite unclear to them what is *added* to the current scientific cosmology by postulating a Designer who is, from a physical scientific point of view, quite redundant. *The psychological function of Creationist and ID explanations, then, is not the 'scientific' one of explaining the physical universe, but the psychological one of rendering it tolerable, meaningful, and 'unpuzzling' to people of certain kinds of temperament or psychological composition.* I must stress as strongly as possible that I am not arguing that something is wrong with them. To do so would be to pathologize the majority of the human race.

Perhaps the heart of the matter is that the religious tend to believe that meanings and values originate from outside the human mind and are thus to be objectively discovered in the external world, or revealed by a divine non-human agency. This, I fear, is not the case. It is we who actively create meanings and values, and we perceive the world in terms of these. This does not, however, mean that they are 'unnatural': only that we are where they are sited in the universe (there may be other locations too, but we have no knowledge of them).

It might be objected here that the puzzle in question is far simpler, namely how to resolve the 'cognitive dissonance' generated by needing to believe that the biblical account of creation is literally true as the Word of God, in the face of a dominant scientific world view which is incompatible with it. There are, however, many Christians who do not feel this 'dissonance', at least in these terms; they are quite reconciled to evolution, the theory of the Big Bang origin of the universe, and much else. This indicates that there is something more at stake for supporters of Creationism and ID than needing to resolve a straightforward conflict between scripture and science.

The argument so far

Let us review the argument so far. In Chapter 2 we saw how one basic appeal of the ID argument rested on a false perception of complexity as signifying design. The case was also made that in fact the physical external universe showed no signs of being designed in any ordinary sense, since it lacked any identifiable purpose, and there was no evidence of any external agency operating on it in the way that human designers do on their raw materials. However, even if this was accepted, the fact remained that this self-running universe still operates according to very complex and specific natural laws. Had these been in any way different, we now believe that the universe would have long ago collapsed or evaporated. Surely, it might be argued, these laws themselves signify the involvement at some stage of an intelligent creator?

In this chapter we have seen that the concept of 'law', like that of 'design', is being used in a misleading analogical sense. Natural laws cannot be broken and are inherent in matter itself, not imposed on it from outside. What we might equally reasonably conclude is that they are more akin to the laws of mathematics and logic, and that while the origin of the universe might remain a mystery, *if* any universe is to exist at all it simply *has* to exhibit these 'laws' as a matter of logical necessity. Moreover, these 'laws' are really no more than succinct human-created formulæ for summarizing empirically identified regularities regarding how various forces and substances relate to one another. There are no 'laws' as such in nature, over and above its contents, any more than there are numbers as such.

We then considered the concept of 'cause', considering first Aristotle's concept of 'final causes' (the others in his famous quartet being 'material', 'efficient', and 'formal'). Clearly, the ID case rests on the need to invoke final – and to some extent formal – causes to explain the physical universe, in addition to the first two. But on closer scrutiny we find that these are not such distinct categories as Aristotle assumed, and that scientists have, on the whole, been quite successful in eliminating them from their explanations of the physical world by showing how they can be assimilated into 'material' and 'efficient' cause terms. At first sight this is rather depressingly reductionist and leads to a crude determinism. For a long time many scientists themselves accepted this, either with stoical resignation or with mischievous glee, while some shared the gloom and sought ways round it. There are still those in each category. There are also important philosophical issues which raise doubts about the conceptual validity of the simplistic opposition of 'determinism' to 'free will'. For our purposes, on examining the concept of 'cause' more closely, and relating it to the nature of 'explanation', it emerged that things are far less straightforward than that strict deterministic image suggests. We have been misled by a tunnel-vision notion of material and efficient cause–effect sequences as single linear chains, rather than appreciating that in the real world vast numbers of these, of many different kinds, are constantly interacting in 'real time'. A single event, object, or phenomenon may be 'explained', potentially, in terms of numerous cause–effect chains. Which one we opt for depends entirely on why we want the explanation, on what is puzzling about it to us.

In its entirety, so to speak, any object, phenomenon, or event is always at a unique confluence or intersection point of 'causes' and happens at a specific, unrepeatable, moment in time.

While not entirely exorcising the demon of determinism, this clearly alleviates some of its oppressiveness, for it leaves the future open to unforeseeable, unpredictable, combinations of circumstances and confluences of causes, the results of which are quite unknowable, even though in a strict sense they remain 'determined'. (Exploring this further would get us into the realms of Chaos and Complexity Theory, mentioned before, and the nature of probability.) Considering the nature of explanations further, we noted that their validity depends on how well they succeed in leaving the person requesting them no longer puzzled. Sometimes this involves demonstrating to the enquirer that his or her puzzle was somehow invalid in the first place, being based on a premise that was itself false.

In relation to our topic, then, we have to ask what the puzzle is that ID is intended to resolve. Its supporters present it as a kind of empirical or scientific theory or hypothesis, yet within the framework of current science, and bearing in mind the points that we have been making so far, there is no obvious need for it. Science is not coming up against puzzles which it has reason to believe will not be resolvable using its existing, remarkably successful, approaches. It continues, with good reason, to assume that resolving its 'scientific' puzzles about the material and efficient causes of things will continue to pay off. Even ID's adherents are hard pressed to specify what the scientific benefits of adopting it would be in actually *advancing* scientific knowledge. Finally, then, it was suggested that the kind of puzzle that ID is really trying to remove is of a rather different kind altogether. It is the age-old and universal puzzle of how to find human meaning in an apparently totally indifferent universe – and ironically it is this very *absence* of obvious meaning which renders the need to find it so intense. It is why we construe disasters as punishments, and good fortune as a reward. Biblically, the most powerful attempt to address this is the *Book of Job*. It is also perhaps why popular images and concepts of God have invariably been so contradictory: on the one hand He loves us and cares for us, on the other He may become, in some contexts, viewed as so emotionally touchy that He will unhesitatingly massacre us in our thousands if we inadvertently upset Him.

At this point I should reiterate that I am not here intending to assail religion or dismiss the God concept in general. There are some genuine issues which will be addressed in a later chapter. My chosen remit is only the much narrower one of elucidating the fallacies underlying ID and fundamentalist Creationism. So far I have been concentrating on the concept of Intelligent Design. What I hope I have shown is that it rests on a variety of errors: erroneous analogising, or extension beyond their meaningful use, of the concepts of 'design' and 'law', an erroneous classification of itself as an empirical 'scientific' hypothesis or theory, and a failure to confront the real nature of the puzzle that motivates it. It has also over-relied on people's incorrigible willingness to believe that what is beyond their personal comprehension must be supernatural, and their readiness to conflate any notion that the universe requires an originating agency with the need for 'God' as popularly conceived in one of the major monotheistic religious traditions.

I now want to look more closely at the 'science' question: in particular, turning from cosmic origins to the history of life on Earth, at the notion that evolution is 'just a theory'.

4 Evolution: 'just a theory'?

Nothing is more guaranteed to raise scientific hackles than the phrase 'It's just a theory ...', which is so routinely applied by opponents of evolutionary theory. The reason for this reaction is that it involves a sloppy equation of 'theory' to 'speculation'. In addition, it is invariably followed by the phrase '... not a proven fact', which implies that 'facts' are theory-free, incontestable, objective things. This too is something which no contemporary scientist or philosopher would accept. A scientific theory is *not* just a speculation, and 'facts' are not necessarily independent of theories; indeed, it is our theories which provide us with the concepts that we use to identify, understand, and define the facts. The word 'malaria', for example, means 'bad air', because originally the disease was believed to be caused by odious and mephitic vapours rising from stagnant water. 'Influenza' was so called because it was explained astrologically as due to some malign stellar influence. Clearly, then, nobody has, strictly speaking, ever suffered from malaria or influenza, and we can continue to use these terms in everyday language only because we have forgotten their etymological roots. This is not mere pedantic logic-chopping. What we understand the facts of nature to be is determined by our 'theory' of the world. There is no theoretically neutral way of stating them. Even calling a creature a 'bird' assumes the validity of the Western zoological classification; other cultures may have systems which include bats and dragonflies in their term for flying creatures (and exclude ostriches). One could even go so far as to observe that saying one thing is 'above' or 'below' another implicitly assumes a flat Earth. There is no 'objective' 'up' or 'down' in relation to the Earth's position in space or the surface of a globe. More scientifically accurate would be statements about proximity to the Earth's centre of gravity – 'above' being more distant, and 'below' being closer. Naturally, there is no practical need to introduce this innovation in daily life, although wider circulation of maps with South at the top might be beneficial.

While philosophers still find much to debate concerning the precise meaning of the word 'fact', none would accept the crude *theory vs. fact* dichotomy as valid. One interesting observation on the fickle nature of 'facts' was made by the philosopher John Austin in *How to Do Things with Words* (1962). He pointed out how the truth or falsity of a proposition often depends on the purpose that it was meant to serve: thus for some purposes it is quite acceptable to say 'The distance from London to Brighton is 60 miles', but for others this is far too imprecise, and in any case one would have to specify the precise locations in London and Brighton from which the measurement was being taken. This would draw us into quasi-theoretical considerations about how the borders of London and Brighton were to be defined, and where their 'centres' are located. And when trying to measure the length of a coastline, precision becomes literally impossible below a certain level. Again, is it true that North America is triangular in shape? Well, yes for some purposes – and no for others. So perhaps there is a lesson here about propositions such as 'God created the universe'.

Before proceeding, I should note that the model of science that I am deploying here will strike many more sophisticated readers as rather simplistic. But to attempt to go further would, in this context, complicate things unnecessarily. The status and nature of 'scientific' knowledge has been a matter of profound debate among philosophers of science, sociologists of knowledge, and historians of science since the mid-twentieth century. From the work of philosophers of science such as Thomas Kuhn, Norwood Hanson, Imre Lakatos, and Paul Feyerabend in the 1950s and 1960s, through Bruno Latour onwards, many questions have been raised regarding how such knowledge is created and evaluated, and how far these processes should be considered strictly logical, as opposed to simply 'rational' in some less formal sense. Issues related to the 'social construction' of knowledge and, indeed, of the objects of knowledge, have been, and remain, matters of intense debate. I am well aware of these issues, and it should not be assumed that I am personally espousing either a particularly strong 'positivist' position or a 'realist' position in what follows. Most would in fact classify my position in previously published work as broadly 'social constructionist', although 'historical constructionist' or 'mutualist' would be more accurate labels. However, the fact that these critical questions *can* be raised regarding science should not be read as offering any comfort to Creationists and the ID camp. 'Knowledge' of God, and religious 'facts', are even more demonstrably 'historically constructed' and vulnerable to radical, socially contingent change over time than the 'knowledge' offered by the physical sciences. One would be hard put to find a single 'fact' of religious doctrine that is as indisputably worthy of a place in Bruno Latour's closed 'black box' category (i.e., no longer practically worthwhile challenging) as are the chemical 'periodic table' and the tenets of basic genetics, for example.[1]

Criteria for evaluating scientific theories

Turning from facts to theories, we should immediately note that there is a broad consensus in the physical sciences that for practical purposes theories have to meet a number of criteria to be accepted by the scientific community. The list presented below is a broad summary of the criteria conventionally accepted by orthodox physical scientists since Karl Popper's *The Logic of Scientific Discovery* (1959), which originally appeared in German as *Logik der Forschung* in 1934. Since then, numerous philosophers and historians of science have challenged aspects of Popper's position. There are genuine issues regarding how accurately these criteria reflect real scientific practice, and whether they are all fully sustainable in detail, but they can still serve as an initial reference point in representing how most scientists, if not all philosophers of science, view the matter. In any case, as noted above, the deeper philosophical shortcomings of this list offer little consolation to Creationism and ID. They are also harder to apply to the human sciences such as psychology and sociology, within which there is far less of a consensus.

The criteria that scientific theories should meet may be summarized as follows.

- They must account, more comprehensively than alternatives, and in as economical and logically consistent a fashion as possible, for the known data regarding the phenomena with which they are concerned.

- They should be consistent with currently accepted theories in other disciplines; if not, then either or both will require modification or rejection (see below). The scientific account of the universe is now so complex that theories regarding different phenomena, the subject matters of different disciplines, have come to interlock and mutually reinforce each other quite inextricably.

- They should be able to predict future data, yielding hypotheses which can be empirically tested. In a nutshell, they should be *productive* in generating knowledge. The sheer number of things whose very existence has been established by scientists pursuing their theories is enormous, from vitamins, viruses, hormones, and proteins to neutron stars, photons, electrons, and X-rays. And 'facts' about these can be stated only within the conceptual framework of the theory that brought them to light. When such a theory fails or is refuted, the 'facts' about these things must, in turn, be reformulated, as the fate of 'phlogiston' in chemistry famously illustrates. 'Phlogiston' was a gas proposed by eighteenth-century chemists to explain combustion. It had, among other things, the curious property of negative weight, because 'obviously' weight was lost during the combustion process. Phlogiston was a normal component of air, and combustion in a closed space left the air 'dephlogisticated'. The discovery of oxygen laid phlogiston to rest, but it would be a mistake to say that phlogiston was an earlier name for oxygen, since its supposed properties and mode of operation were quite different. This closely relates to the next feature.

- It should be possible to specify what would count as falsifying or inconsistent data. When this is found, it may prove possible to adjust or amend the theory, or the theory may fall. How apparently falsifying or inconsistent findings are actually handled depends very much on the specifics of the case, and the point at which a theory must be considered falsified is in practice largely a matter of peer consensus. That such a point should be identifiable is nevertheless essential.

• And in practice, for a wide range of disciplines, though not for those such as astronomy dealing with matters beyond any human control, we also expect good theories to have practical consequences, such as facilitating new technologies and new medical treatments.

In principle, scientific theories are always open to replacement or revision in the light of new research findings and discoveries. In reality, of course, some, like the theory that the Earth goes round the Sun, not vice versa, have become so fundamental that their future rejection in this way is to all intents and purposes inconceivable (as mentioned before, Bruno Latour calls these 'black boxes').

Evolutionary theory meets these criteria fairly well. The legitimate quibble may be with whether it is a single theory at all, or a conceptual orientation within which a variety of more specific theories may be developed. There are indeed some quite central on-going controversies among evolutionary biologists, particularly between hard-line sociobiologists like E.O. Wilson and Richard Dawkins, and 'punctuated-equilibrium' theorists such as the late Steven J. Gould, R.C. Lewontin, and N. Eldredge. The situation is further complicated by how far classical 'natural selection' needs to be, or can be, supplemented by Complexity Theory principles and the operation of 'laws of form', as argued by Brian Goodwin in *How the Leopard Changed its Spots: the Evolution of Complexity* (1994), or whether 'group selection' needs to be invoked. Most crucially the Gould camp strongly argue that while natural selection is one very important evolutionary mechanism, there are other, less obviously 'adaptationist' processes in play. For a very accessible summary of criticisms of strong 'adaptationism', Steven Jay Gould's 'More Things in Heaven and Earth' (2000) is most helpful.[2]

Even more spectacularly, since about 1990 there has been a growing body of evidence for the phenomenon of 'horizontal gene transfer' (HGT). Instead of life forms evolving in simple lineages, with genes being passed from one generation to the next, it now appears that they can be passed horizontally across often widely different species. This appears to be ubiquitous among micro-organisms such as bacteria (which actually comprise the majority of all known species and in terms of tonnage contribute 90 per cent of the total weight of living organisms on our planet). It also happens in more complex organisms up to, and including, humans. It is early days for research in this field, and quite how the transfer is mediated is not entirely clear, as several different mechanisms appear to be in play. One which appears very important is via viruses. The role of HGT in evolution is now being hotly debated, but one consequence is that the familiar traditional image of evolution as resembling a branching tree is fast becoming inadequate. In recent decades it has already come to resemble a tangled bush rather than a tree, and now HGT is suggesting something more like a web.

Does this falsify evolution as a theory? Actually no; but what it does do is suggest that the raw material of variation on which selection works can occur by means other than genetic mutation.[3] There is little comfort here for Creationists or ID proponents. Moreover, the very fact that such an initially heretical notion as HGT has been able to develop and flourish demonstrates that evolutionary scientists are far from dogmatic and are disinclined to suppress new ideas.

I would note here that students, hankering after concrete facts, often think that the existence of controversies like these means that 'nobody knows', so 'it's just a matter of opinion' – thereby signifying disciplinary failure or weakness. On the contrary, they are the very life-blood of active fields of knowledge, scientific or otherwise, signifying where the action is. Facts beyond dispute may be essential to know but are dead legacies of past controversies. A student unable to grasp this point might as well give up.

In relation to this point, it is worth putting evolutionary theory into historical perspective here, especially as it pertains to human evolution. Scientists have been studying evolution for only about 200 years, or barely 150 if you wish to take Charles Darwin's 1859 *On the Origin of Species* as the starting point. If continuing disagreement is a sign of failure, consider the state of Christianity at a similar point after its inception. A Victorian scholar, the Revd. James Henry Blunt, identified about 50 'heresies' within Christianity by 200 C.E. in his 1874 *Dictionary of Sects, Heresies, Ecclesiastical Parties and Schools of Religious Thought*, while the New Testament as we know it had yet to be compiled by 200 C.E. Scientific research into human evolution really dates only from the 1860s at the earliest – hardly 150 years (or about five generations of researchers) ago. Even by 1960 the number of fully qualified scientists actively engaged in this research had in all likelihood not yet entered three figures (although I stand to be corrected). In terms of funding, it was far less affluent than other fields with clear applied military, industrial, or medical pay-offs. Events such as the recent discovery of apparently very early hominid fossils in Georgia (Europe, not the US state!), which apparently raise problems for the standard 'out of Africa' theory, cause excitement and rethinking among human-evolution researchers, not groans of disappointment and fear that the entire evolutionary approach has been called into question. Despite numerous theoretical disagreements and debates, those studying evolution have found nothing to challenge the fundamental premise that there is an evolutionary process underlying the emergence of all life-forms. By comparison, Christianity around 200 C.E. was riven by often bloody disputes over doctrinal essentials such as the nature of Christ's divinity.

One irony (of many) in the current context is that the resurgence of Creationism/ID could be having an inhibiting effect on scientific debate, as evolutionists of all kinds may feel it necessary to close ranks and present a solid front. Most scientists would now, I think, consider evolution to be a conceptual orientation rather than a single theory; nevertheless it meets the criteria described above impressively well. Let us take them in turn.

- We have yet to encounter data which evolution cannot, in principle at least, account for. Since Darwin it has undergone numerous modifications, and the mechanisms involved have been elaborated in much greater detail. Although our current understanding of genetics has entailed some rethinking of the dynamics by which it is played out, the broad notion of 'Natural Selection' remains both logically simple and largely sufficient, even though the adequacy of the simple 'adaptationism' which this implies has, as just mentioned, been very seriously challenged by writers such as S. J. Gould. For what it's worth, my personal sympathies are with the critics. To see these doubts about the full sufficiency of natural selection as providing loop-holes for ID would, however, be mistaken: to invoke a 'God of the Loop-holes' would be as foolish as invoking a 'God of the Gaps' once proved.

- Evolution is consistent with the theories currently held by scientists in other fields. For example, its time-scales are reinforced by dating methods based on nuclear physics, and it is consistent with and illuminated by genetics.[4]

- Evolutionary predictions about relationships between various life-forms have been confirmed by measures of DNA differences.

- While evolution has proved flexible enough to incorporate data which have at various times seemed anomalous, it would quite obviously be in serious trouble if radical inconsistencies with genetics or palæontological (fossil) evidence emerged. I should stress, however, that the weight of supporting evidence is now so great that a handful of anomalous fossils would be quite insufficient to bring the theory of evolution tumbling down, unless their discovery and geological context had been rigorously documented and scrutinized in accordance with accepted palæontological standards. Since the famous 'Piltdown Man' forgery in the early twentieth century, palæontology has, with good reason, become ranked among the more paranoid of scientific disciplines when it comes to amateur discovery claims. (See Appendix A for a brief discussion of the nature of fossil evidence.) Gaps in the fossil record are not themselves counter-evidence – as the saying has it, 'absence of evidence is not evidence of absence' – and most critics of evolution long ago abandoned fossil gap-spotting as a tactic.

- Regarding practical applications, however, these are less clear by the very nature of the subject. Some would argue that evolutionary understanding of the origins of certain human behaviours can be of benefit to psychologists and psychotherapists. More broadly the notion that humans evolved from primates may also be judged to have had salutary effects on how we view ourselves and our relationship with the natural world, while the evolutionary approach in general has aided the rise of ecological understanding. There have, however, in the human sciences and in culture generally, been some far darker episodes, especially in the first half of the past century, when evolutionary thinking was used as a rationale for evil events. Eugenics, Degenerationism, and 'Scientific Racism', which were eventually adopted by the Nazis, provide an enduring cautionary moral lesson for all evolutionists who seek confidently to apply its concepts to contemporary human affairs.[5] None of these 'applications' of evolutionary thought is really akin to the technological applications of other physical sciences (although eugenic sterilization of the unfit is close to being an exception), and their occurrence has no bearing on the validity of evolutionism as a scientific doctrine. Opponents of evolutionary theory, both Christian and Islamic, have sought to make much rhetorical use of them, but this is like arguing that medieval witch-burning refutes Christianity, or that the events of 9.11 disprove Islam. One should indeed be wary of attempts to apply evolutionary concepts to contemporary social affairs, not necessarily in principle but because their past misuses show how easily they can be co-opted for non-scientific ideological agendas. What is crucial is that the evolutionary perspective is integral to the present scientific world-view and cannot be extracted from it like a sore tooth without disrupting it in its entirety.

Turning now to the Creationist alternative to evolution, in the light of these criteria it has to be said that it does not fare well at all. Creationism, as we have seen, comes in two versions. Strong Creationist thought will concern us in more detail in Chapter 6; here we will confine ourselves to indicating Creationism's general shortcomings as a scientific theory.

- It must be admitted that if one accepts the premise of biblical infallibility it does indeed account for all phenomena in a simple fashion: 'God created it all'. The Genesis account is so brief and lacking in detail, so schematic, that it is easy to fit everything into it. But that is as far as it goes, and the Genesis account is hardly logically coherent (see Appendix B).

- Is it consistent with all the other scientific theories? Well, as it was written before any scientific theories in the modern sense were thought of, it is difficult to see in what sense the Book of Genesis relates to them at all. Even leaving aside the literal claim of 'six days', such few 'empirical' propositions as it appears to contain do not, on the face of it, seem to be consistent with the scientific theories currently accepted as relating to the topics in question (even the schematic order in which life-forms appeared). I might note here that *Genesis* 5 must also be considered quite anomalous, listing as it does a succession of Adam's immediate descendants from Seth, via Enos, Cainan, Mahaleel, Jared, Enoch, Methuselah, and Lamech to Noah, who all lived for more than 800 years, except Lamech (777) and the unaccountably short-lived Enoch (365), Methuselah famously coming top at 969. (Noah's death at 950 does not come until Chapter 9.) Are we really supposed to take these statements literally? If so, then this would certainly be inconsistent with everything we know about human biology and the ageing process. The Flood had an immediate limiting effect on such longevity: Noah's son Shem seems to have made it to 600 (*Genesis* 11, 10–11) and his descendants managed the 200–400 year range until Abram's grandfather Nahor, who died at just 148 (*ibid.* 24–25) although his father, Terah, made it up to 205 again (*ibid.* 32). These figures are important, because ever since Bishop Ussher they have continued to provide the 'empirical' basis for literalist calculations of the age of the Earth.

- When we come to its ability to generate testable hypotheses and discover new facts, Creationism fails completely. I honestly do not know of a single discovery resulting from 'Creation Science' or testing ID-generated hypotheses. And its practical technological pay-offs have been notable by their absence. Its exponents have, moreover, never made clear what it is they would accept as falsifying it.

What I am saying here echoes and elaborates on the points made by District Court Judge John E. Jones III in December 2005, in rejecting the insistence of the Pennsylvania Dover Area School Board that ID be compulsorily included in the school curriculum. I quote from the summary provided in George J. Annas (2006):

...the expert testimony demonstrated to him that intelligent design is 'an interesting theological argument' but is not science for many reasons: it invokes a supernatural cause; it relies on the same flawed arguments as creationism; its attacks on evolution have been refuted by the scientific community; it has failed to gain acceptance in the scientific community; it has not generated any peer-reviewed publications; and it has not been the subject of testing or research. The judge quoted from a report on creationism by the National Academy of Sciences as an authoritative and definitive source: 'Creationism, intelligent design, and other claims of supernatural intervention in the origin of life or species are not science because they are not testable by the methods of sciences. These claims subordinate observed data to statements based on authority, revelation, or religious belief.'

ID and Creationism can in fact be viewed as parasitic, since, while signally failing to generate any concrete knowledge themselves, their proponents constantly harry the physical sciences by claiming, frequently on the basis of rumour, to have found fatal, 'falsifying', flaws in its accounts of geology, evolution, or even fundamental physics (for example, the red shift is caused by light 'ageing' and is not an index of the distance of its source – an interpretation which is, to put it charitably, staggeringly muddle-headed). The fact that none of these claims has withstood serious scrutiny in no way dampens their proponents' ardour for the game. But the game is not 'science', not even the 'falsification' aspect of it. Scientific falsification must itself be based on established procedures to convince scientists (and these may vary between disciplines to some extent), and as far as I am aware the cases invoked very rarely indeed emerge from following such procedures. More often, like the enigmatic 'Coso Artefact' allegedly found inside a geode, they will turn out to be a 1920s Champion A spark plug in a lump of heavily impacted clay (this hilarious story is told in detail on the www.talkorigins.org website).

One Creationist/ID criticism of evolutionary theory, or 'Darwinism' as Creationist Americans still prefer to call it, which does demand a little attention is the claim that it is 'tautological'. It is indeed true, regarding the doctrine of the 'survival of the fittest' (incidentally, a phrase coined not by Darwin himself but by his contemporary the evolutionist Herbert Spencer), that almost by definition 'fittest' – meaning best-adapted to the environment – and 'survival' – meaning able to live long enough to reproduce – are virtually synonymous, since 'adapted to the environment' can really only be defined as 'able to live long enough to reproduce'. The mechanism of 'natural selection' may also be viewed as tautological, since it means only that organisms which do not live long enough to reproduce will have no offspring, and those that do will – hence, insofar as their survival was due to inherited characteristics, they will pass these on to such offspring.[6] One possible problem for evolution's opponents is that if these are tautologies they are necessarily true by definition! But are tautologies valueless? Very often they are, but what they can do is focus attention on important, but obscure, conceptual truths by bringing different conceptual frameworks together and integrating them. In the case of evolution, they brought together concepts regarding population levels ('demographics') and the nature of heredity ('genetics', as we now call it) in relation to the phenomenon of the diversity of organic life. This may seem obvious now, but it was not so 150 years ago.

These 'empty tautologies' thus generated an intense empirical scientific project of unravelling precisely how heredity worked, how inherited traits enabled organisms to survive and reproduce, and how environmental factors acted on organisms in a way which affected their reproductive chances. They are not themselves 'the theory of evolution' in its entirety, but simply its historical starting point or, if you like, axioms akin to those of Euclidean geometry.

Incidentally, the US custom of calling evolutionary theory 'Darwinism' is an interesting rhetorical ploy. It misleadingly identifies modern evolutionary theory as purely the product of Charles Darwin, when in fact it assumed something like its current form only in the 1930s with the work of R.A. Fisher, Sewell Wright, and Julian Huxley, and has continued to develop and change ever since. Even in the late nineteenth century, Darwin's version was effectively eclipsed after around 1880, evolutionists much preferring to focus on mechanisms other than natural selection. Darwin himself, indeed, believed that 'natural selection', while the most important mechanism, needed to be supplemented by 'sexual selection' and, possibly, Lamarckian mechanisms ('inheritance of acquired characteristics'). To explore this further, readers might well start with Peter Bowler's two illuminating books: *The Eclipse Of Darwinism. Anti-Darwinian Evolution Theories in the Decades around 1900* (1983) and *The Non-Darwinian Revolution. Reinterpreting a Historical Myth* (1988). Calling it 'Darwinism' has two further subtle effects: (a) it implies a dogmatic adherence to the first version proposed by a long-dead 'founder', which is quite erroneous; and (b) in associating it with this single figure it hints that it is cult-like or ideological, like 'Marxism', and thus can be marginalized from the rest of science. Evolutionists do not constantly pore over the words of Darwin seeking enlightenment on current questions. *The Origin of Species* is not a religious text.

    To return to evolution: the evolutionary account of the history of life on Earth is far from complete, and the mechanisms by which evolution operates are not yet fully understood. The core tenet of evolutionary thought is not, ultimately, that evolution was driven by any one mechanism, but that life on Earth originated, and subsequently developed, as a result of the operation of mechanisms which are, or will be, explicable by orthodox science, at no point requiring the intervention of an external divine agency to take the course that it did, up to and including the emergence of humankind. The intervention of non-divine extra-terrestrial aliens is another matter. While nobody currently takes this hypothesis seriously, there is nothing inherently 'unscientific' about it, and one could imagine evidence turning up which might support it. This distinction, I note, does not appear to figure, or is at any rate blurred, in Stephen C. Meyer's pro-ID paper entitled 'The Scientific Status of Intelligent Design: the Methodological Equivalence of Naturalistic and Non-naturalistic Origins Theories', published in the collection *Science and Evidence for Design in the Universe* (2002). Over time the parameters for uncertainty in how this happened get progressively narrower, particularly as our understanding of genetics continues to grow and our ecological vision of how life-forms interact becomes more sophisticated. For a full statement of the nature of evolutionary theory, the reader is referred to Stephen Jay Gould's massive, posthumously published, *The Structure of Evolutionary Theory* (2002) (although this appeared too early for HGT to be covered). An equally solid account of Creationist or ID theory would be an intriguing document indeed.

The popular works by Richard Dawkins, notably *The Blind Watch-Maker* (1988), are also useful up to a point, but one does feel that his scientific agenda is overlaid by a more personal, emotionally rooted, philosophical one which rather affects the register of his discourse. His famous concept of the 'selfish gene' is a particularly sloppy rhetorical move, and the use that he makes of it emerges as a kind of covert dualism in which some agency, 'my genes', affects 'me' as a separate, presumably immaterial, entity. Philosophically this just will not do. His pro-evolutionary fervour should not mislead readers into thinking that his particular vision of evolution and his attitude towards it reflect a scientific consensus. He is actually falling, in these works, between the two stools of 'scientific' and 'philosophy of life' styles of discourse (to be discussed later). For many anti-Creationists his later work, *The God Delusion* (2006), prompts the question 'With friends like these who needs enemies?'.

There is, then, nothing in Creation Science, as far as I can see, which could aid genetics, ecology, or any other relevant field of scientific research. Rather, it wishes to call a halt to the entire proceedings and convince us that the conviction that 'God created it' will satisfy all our curiosity and resolve all our puzzles.

The evolutionary picture

A very brief recapitulation of the core evolutionary processes may be in order here, bearing in mind what was said about complexity in Chapter 2. Quite how the very first single-celled organism originated remains indeed obscure, but there are numerous scientific hypotheses in play. Once a life-form *had* come into existence, the evolutionary game was nevertheless afoot. In order to survive, life-forms need, as an absolute minimum, nutrition. Initially this would, necessarily, have been inorganic, and as long as it was plentiful the population could happily expand. The crunch comes when nutritional resources become scarce: when demand exceeds supply. There are three ways of responding to this: become better at accessing them, become more organically efficient at using them, or switch to something else. And now the evolutionary process really kicks in. Life-forms begin to variegate as chance genetic mutations enable them to respond in one or other of these ways. Being microscopic and reproducing quite rapidly, simply by dividing (as amoebae do), there were soon billions of individuals in this initial population, so such mutations would hardly have been uncommon.

From then on, another factor comes into play: the waste products of the life-forms, consisting of gases and minerals, gradually change the environment in which they are living, and further adaptive adjustments are required. Extraneous geological events and climate changes, even meteor impacts, also enter the picture. As organisms adapt to the changing conditions, many become more complex, with new features being integrated into inherited ones. A remnant of the ancestral population may well endure more or less unchanged in favourable environmental niches for a long time, but the broader environmental changes now in train will usually eliminate them eventually. Nor of course will any genotype remain entirely unaltered, because perfect DNA replication cannot be sustained indefinitely. (Horizontal gene transfer, mentioned above, is also apparently in play in generating variations on which selection can operate.) Whether changes benefit, harm, or leave unaltered the organism's chances of survival is a matter of luck, but evolution requires only a few to fall into the first category. Organic life has major effects at a geological level too, resulting in chalk and the 'fossil fuels' of coal and oil. The composition of the atmosphere also shifts with respect to its oxygen, nitrogen, and carbon-dioxide ratios.

The essential point to grasp here is that, given the immense time-scales involved and our current knowledge of genetics and geology, the mechanisms indicated above provide the basis from which a quite satisfactory, if still provisional, scientific understanding of the material and efficient origins of today's planetary biosphere can be achieved. And contemporary genetic and geological knowledge is itself in large part a result of the stimulus that 'the Theory of Evolution' provided a century and a half ago. Of course there are still aspects of this which remain puzzling, and the adequacy of the strong 'adaptationist' 'natural selection' model has been very plausibly challenged, as previously noted. Actually, as we have seen, Darwin himself, in the Conclusion to the sixth edition of *Origin of Species* (1872), wrote that natural selection was the 'main but not the exclusive means of modification' (p. 421). He also identified 'sexual selection' as a distinct sub-variety of natural selection. His exposition of this came in the second half of the book, usually referred to simply by its main title, *The Descent of Man* (1871). It refers to the role played by mate-selection in determining reproductive chances, and he argued that it accounted for the more exotic features, such as elaborate plumage and huge horns, of males, and sometimes females, in many species. In other respects these features are often quite useless, or even dysfunctional. They are used either to render one party more attractive to the other (plumage, for example) or to achieve intra-gender dominance and access to females (horns, for instance). Such features are typically associated with behavioural 'dance' rituals. When humans too find the feature attractive, the species is in trouble and likely to be slaughtered in large numbers. Acquiring possession of such body parts presumably now enhances the sexual attractiveness of their human owners.

The main blind-spots in earlier evolutionary thinking were, I feel, a failure to appreciate the dynamic ecological interconnectedness of the process as involving multi-level feedback mechanisms, and the effect of fortuitous catastrophic events such as meteor impacts and geological upheavals. Lack of a clear grasp of the sheer complexities and subtleties of genetics has also often limited theoretical sophistication, and possibly still does. These limitations do not, however, provide Creationism and ID with much comfort, reinforcing as they do the image of the biosphere's complexity resulting unintentionally from the interactions of its component agents, both with each other and with the non-biological environment, as they each pursued their own immediate survival interests. Personally I do have some sympathy for James Lovelock's holistic 'Gaia hypothesis', mentioned earlier, in which the current biosphere is viewed as a kind of self-regulating 'meta-organism', but this is irrelevant to the immediate argument.

Two final thoughts

The first of my final thoughts at this point is this: the idea that Nature displays, in its complexity and beauty, the wisdom of a divine creator has not been unambiguously shared by Christians of all generations or denominations, although it was central to the appeal of the Argument from Design around 1800. Nature's obvious utter indifference to humanity's interests, demonstrated in flood, earthquake, famine, disease, and volcanic eruption, led many in the past to see the Earthly realm as belonging to the Devil, a vale of sorrow, controlled by the Enemy, through which we must pass. Faith in Christ was faith in His ability to redeem us and bring us into the superior divine realm – not in His ability to do much about our Earthly travails – other than enable us, via His love, to endure them. The Earth was also seen as Fallen: 'Change and decay in all around I see', as the famous hymn 'Abide with Me' has it. In presenting Creationism as representing some kind of traditional core Christian consensus, Creationists are, I fear, displaying their ignorance of history. And if they want to argue *both* that the beautiful harmonious complexity of the universe implies a creator *and* that He has to intervene because we now live in a Fallen disharmonious world, they would, as said before, really be trying to have it both ways.

My second thought is more serious. Any moral virtue that religious belief possesses, especially Christianity, is generally held by believers to lie in the fact that the belief was freely chosen. Some notion of free will is absolutely essential. The notion of choice is at the very core of the traditional Christian world view. It is *we* who must decide whether to accept Christ into our lives and submit to God's will. But the Creation Science and Intelligent Design projects run in the diametrically opposite direction. In striving to prove 'scientifically' that the fundamentalist Christian world view is factually and demonstrably correct, they are missing the whole point. Success would remove virtuous choice altogether. We could no more *not* believe than we could doubt that 2 + 2 = 4. It is crucial that open-minded Christian readers in particular grasp this, naturally tempted as they may well be to sympathize with ID and Creationism. Creationism's proponents are, in short, unwittingly striving to saw through the very branch astride which they sit. Mercifully perhaps, as far as most scientists are concerned, they have yet to penetrate the bark. Eliminate free will and you remove the heart from your religion.

In Chapter 6 we will explore further some of the psychological features and traits of Strong Creationism, features which Creationists of a less literalist kind, along with proponents of Intelligent Design, share to varying degrees. But before that I wish to clarify a fundamental error which proponents of Bible-based Creation Science display regarding the nature of the Bible. They appear to view it as some kind of 'scientific' text, when in fact, as a traditional sacred text, its character and functions are the very opposite of those of modern scientific texts. Insofar as it does contain literal accounts of the nature of the physical world comparable to those offered by science, these are in fact its least valuable content and easily expendable. It is a strange *non sequitur* to claim, as does the leading Creationist Henry Morris, that the 'inerrancy' of the Bible is in effect non-negotiable because if the *Genesis* account of creation is wrong, then it might also be wrong about Christ and salvation.[7] It seems symptomatic of a very high order of certainty-fetishism indeed. How has this error arisen?

5 Why some books endure

In the light of the previous two chapters, advocates of Creationism and its ostensibly less religion-based offspring, ID, are clearly mistaken if they genuinely believe that they are proposing systems of thought and types of explanation in the same category as scientific theories, to which they can provide scientific alternatives. They are not, in a curious historical reversal, playing Galileo to some Scientific Inquisition. They are simply, to use a technical philosophical term, making a 'category mistake'. If the Creationist accounts are in any sense valid, this cannot be validity of the form possessed by scientific accounts. Central to their position is a mistake regarding the nature of the Bible itself. To clarify this, a different angle of approach is needed.

Some written texts continue to be read even though they date back hundreds, occasionally thousands, of years. The vast majority of all writings are rapidly forgotten, but a few endure. Most of the world's great 'sacred' texts, for example, are among the notable survivors. Why? The Bible, the Qur'an, the Vedas, and the Upanishads were initially written in societies where only small elites were literate and in which access to any written texts was rare. Writing and books were awesome things. And there were very few of them. Books were intended to be read aloud – an important point, now easily overlooked. In order to endure, a text therefore had to be able to sustain its meaningfulness and impact through many readings. Although they had heard or read them since childhood, the elderly still had to be able to react to, respond to, or even *perform* them, afresh, finding some new depth, some hitherto unnoticed or unsuspected nuance. (There is a faint echo of this in the way in which some popular songs endure: each generation of singers finds a new way of interpreting the 'standards' which breathes fresh life into them.) *It was their very capacity to enable people to do this with them that identified such texts as special, wise, sacred, or holy.* And before the late seventeenth century in the West (and for longer elsewhere) it was generally therefore accepted that such texts had numerous levels of meaning. They were not, in other words, just simple expositions of literal facts. Had they been so, they could no more have survived lifelong re-readings or re-hearings than the instruction manual for a TV set's remote control, or a list of English monarchs. (Some readers will appreciate that this argument relates closely to the differentiation between *mythos* and *logos* used by Karen Armstrong, which I briefly discuss in Appendix C. The line that I am adopting in this section in fact appears to be very similar to hers.)

What seems to have happened in European cultures from the seventeenth century onwards – as a result of printing, widespread literacy, and the rise of science – is that a division more clearly developed between texts that were imaginative and symbolic (such as poetry) and those intended to be clear, unambiguous, and literal (although some of the latter genres such as law books and histories had long been well established). A key document in this process was Thomas Sprat's insistence in his 1667 *History of the Royal Society* that 'natural philosophical' writings must avoid metaphor and persuasive rhetorical tricks; they should strive to be unambiguous and straightforwardly factual. (This reflected the outcome of the deliberations of a Royal Society committee on improving the English language, established in 1664.) The separation between method/ results/ discussion sections in research reports was also established in order to separate debatable speculative and theoretical content from incontestable 'facts'.

In this context the status of the Bible in particular became problematic. A tension arose between its supposedly 'literal' meaning and its 'symbolic' ones. It must, it was felt, be one or the other. Ironically this has resulted in many pious Christians feeling that they have to insist on its belonging in the former category in order to remain credible or 'true'. A rich, multi-layered text in which folk-history, poetry, tribal record, spiritual wisdom, speculative cosmology, moral injunction, and philosophy (but rarely science!) are all mingled had to become a one-level text. Where it was patently symbolic (as in the books of *Daniel* and *Revelation*), this could only mean that it was written in some kind of code, translatable into literal terms by using the right key (the 'day equals a year' rule, for example). Certain biblical passages, much of *Numbers* being an outstanding example, are, it must be said, very rarely quoted or taken as sermon texts precisely because they so obviously *do* have just one simple literal meaning, which goes to prove the point. In that case (for example, *Numbers* 26.12–62), we in fact often find the ancient Hebrew equivalent of the tribal genealogy intoned in the last episode of the TV dramatization of Alex Hailey's African–American saga *Roots,* when the hero finally returns to the African community from which his kidnapped ancestor Kunta Kinte originated.

Scientific writings, by contrast, are literal texts *par excellence.* They are not read as offering a scale of levels of symbolic spiritual meaning for the questing soul to ascend; they do not seek to serve as comforting texts for young and old alike, nor have they moral teachings to impart by means of allegory or parable. Any enduring value possessed by the world's traditional sacred texts, however, surely lies in *precisely* these features, while their literal, quasi-scientific, passages must be the most easily jettisoned of their contents, being relics of very local pre-scientific understanding of the natural world as it presented itself to their geographically restricted cultures of origin.

The vast majority of primary scientific texts are rarely read by anyone other than their authors' scientific peers, and even then, in today's rapidly moving scientific world, are typically forgotten, uncited, after ten or fifteen years. The works that we cherish and re-read are, with rare exceptions, not scientific but literary, historical, philosophical, or religious texts. We feel this way about them for various reasons. We enjoy the companionship of the author's voice; or they are rich in meaning, entertaining or stimulating in content; or they simply use language in a beautiful fashion. The rare scientific exceptions endure for similar reasons and not for their factual, scientific content. Only historians of science delve into the rest.

From this perspective the efforts of Creationists present themselves as a continuation of those earlier attempts to render the Bible into some sort of scientific text by insisting on its literal factuality in all respects. They have covertly bought into the very idea that 'science' is the supreme ideology which overtly they purport to be opposing, and the notion that only such a status can guarantee its value. But, as I have just tried to indicate, this is the very opposite of the truth. The current value of scientific works is essentially transient, as their contents are constantly being rendered obsolete. It is, to reiterate, poetry, drama, novels, philosophy, history, and religious works which, if of sufficient quality (and favoured perhaps by chance), endure as fresh and meaningful across time. And if biblical quotes are in order, how about St. Paul's statement of the Christian attitude in II

Corinthians iv. 18: '... we look not at the things which are seen, but at the things which are not seen; for the things which are seen are temporal; but the things which are not seen are eternal'? (One can of course selectively quote the Bible in support of just about anything, but being able to do so does seem to remain a necessary condition for being taken seriously by some Christian readers, and this one seems fair enough.) As noted in Chapter 1, the totally literalist mode of reading the Bible was in fact very much a nineteenth-century development. But if its literal meaning is straightforward and transparent, why have those who adopt this approach failed to reach a consensus (for example, taken 'literally', the Bible implies that the Earth is flat, but only a handful of even the most fundamentalist literalists any longer believe this, although in the early twentieth century there was a significant minority who did: see C. Garwood, 2007, discussed in the next chapter). It is also hardly news that the Bible is peppered with inconsistencies and contradictions if taken 'literally', while the Christian understanding of the Old Testament has, since early in its history, been that it is symbolically prophetic of the coming of Christ. Equally, devout Orthodox Jews totally reject this interpretation.[1]

The merits of science lie primarily not in its texts but in its products. While, admittedly, both literature and science can change how we see and experience the world, achieving such an effect is the primary aim of literature and but an unintended consequence of the latter, insofar as it is implicitly encouraging us to view the world 'scientifically', or as it reveals new phenomena which then inspire the literary and artistic efforts of non-scientists (and indeed sometimes scientists themselves in their off-duty hours). Admittedly, plenty of scientists have produced popular expositions of their fields of expertise with precisely this kind of intention, but these works are not *themselves* the 'science': the 'science' is located in the technical specialist literature and the actual research practice as such.

The nature and durability of the Bible: additional thoughts

Regarding the endurance of the Bible itself, some further observations are in order. In the Protestant English-speaking world it is undoubtedly the

King James 'Authorized Version' of the *Holy Bible* (1611) which has served as the most influential version (the Revised Version did not appear until the 1880s). It is of some interest, then, to look at the Preface, entitled 'Translators to the Reader'. This is little known to most readers, since it was subsequently dropped. (I have not ascertained precisely when it was decided to omit it, but I have seen two editions dating from the mid-eighteenth century, from Oxford and Cambridge, which certainly lack it.) This text is not without significance. Two very relevant points immediately emerge from it. The first is that the translators were very conscious of the difficulties of translation and the flaws in all previous translations from the Hebrew and Greek originals, including the very earliest, and they do not exempt themselves as invulnerable to the same difficulties. They explicitly refer, for example, to the problems entailed in translating the Old Testament, in which some ancient Hebrew words occur only once and others, though appearing more than once, clearly have different meanings in different contexts. For this reason the text includes marginal notes ('hanging shoulders', as they are sometimes termed in the printing trade). The translators are also very conscious of the need to avoid obscure words, noting 'No cause ... why the word translated should bee denied to the world, or forbidden to be currant (*sic*), notwithstanding that some imperfections and blemishes may be noted in the setting foorth of it'. They also state that

> wee have on the one side avoided the scrupulositie of the Puritanes, who leave the old Ecclesiasticall words, and betake them to others, as when they put washing for Baptisme, and Congregation in stead of Church; as also on the other side we have shunned the obscuritie of the Papists, in their Azimes, Tunike, Rational, Holocausts, Præpuce, Psyche, and a number of such like, whereof their late Translation is full. ... But we desire that the Scripture may speake like it selfe, as in the language of Canaan, that it may be understood even of the very vulgar.

In fact the translators are quite hostile to linguistic pedantry, fiercely expostulating 'For is the kingdome of God become words or syllables?'

Secondly, and more profoundly, the aim of the King James *Holy Bible* is a political one, as is clear both from the Dedication to King James and, far more so, from the translators' preface. It is to create an official English Protestant version in opposition to the various Catholic versions then in circulation. The entire preface is infused with anti-'Papist' sentiment. What emerged was a literary masterpiece which functioned as the focus for anglophone Protestant Christianity until the 'Revised Version' of the late nineteenth century (which, while perhaps more linguistically accurate and less dated in style, was even at the time considered a step towards blandness). It served, tremendously successfully, as the focus for Protestant understanding of Christianity for almost three centuries. The 'New English Bible' (1961, New Testament; 1970, Old Testament and Apocrypha) only took blandness to a new level ('Emptiness, Emptiness', for example, instead of 'Vanity, Vanity' as the opening of *Ecclesiastes*!).

A more general, perhaps speculative, thought then occurs. Any translation by either an individual or a collaborative collective imposes a particular stylistic tone or 'voice'. In the case of texts like the Bible which are anthologies of writings originally by a variety of authors – and in that case spanning a long time period, two languages, and different locations – the original texts would obviously have had very different authorial 'voices' and writing styles. When translated in its totality into another single language, this variety becomes largely lost and the style homogenized. And of none is this truer than the King James Version. Moreover, as a collective, rather than individual, product, idiosyncratic personal styles and preferences are eliminated, leaving a single impersonal voice with seemingly official authority. This tendency was further reinforced by the archaic nature that the style rapidly acquired (indeed, it was slightly archaic even when first published). This has the effect of further rhetorically reinforcing its authority (although I do not mean that this was consciously intended). So what we have in the King James Version is a literary masterpiece in a resoundingly authoritative style (and, one might add, visual design). If anything could sound like the 'Word of God', it was this. Add the subsequent air-brushing out of the translators' explicit anti-Catholic intentions, and its specific denominational character becomes nigh-on invisible. How much then does its durability in English-speaking cultures owe to these features of the King James Version? Those passages, phrases, and episodes which have entered everyday language have invariably done so as worded in this version.

For many generations of Protestants, this *was* 'the word of God'. But in reality it is but one of a host of translations not only in different languages but emanating from different branches of Christianity. Fetishizing its precise wording as 'literally' God's word is thus highly dubious, to say the least. Moreover, one website that I have visited lists 26 other currently available English translations of the Bible, which can hardly be encouraging for those in quest of its 'literal truth'.

In this context it is highly appropriate to mention, as a final twist in the story, the *Scofield Reference Bible*, a version of the Authorized Version first published in 1909 which has remained popular among Protestant fundamentalists ever since. Edited by Revd. C. I. Scofield with eight 'associate editors', this includes not only an introduction ('To be read'), an index, an atlas, and a guide to Bible study, but most importantly a systematic commentary integrated with the text throughout (thus differentiating it significantly from the usual Bible-commentary genre in which commentaries are published as separate works), along with an updating of the traditional cross-referencing in a central strip between the two columns of text, each strip headed with a date (4004 B.C. for *Genesis* 1) and continued incorporation of the Ussher dates (as referred to in Chapter 1). This is not all, for each book is divided into sub-headed sections (in addition to the usual chapters) indicating the main plot and themes, with preliminary introductions to each book further explaining its meaning and significance. The Introduction expounds Scofield's 'panoramic view' of the Bible as a single unified work, its plot unfolding in five sections: Preparation (the Old Testament – containing four different types of text), Manifestation (the Gospels), Propagation (*The Acts of the Apostles*), Explanation (the Epistles), and Consummation (The Apocalypse, i.e. the *Book of Revelation*). These are closely related to the notion of 'dispensations' – successive stages in God's relationship to Man. The subtitle is worth quoting in full: 'Authorized Version with a new system of connected topical references to all the greater themes of scripture, with annotations, revised marginal renderings, summaries, definitions, chronology and index, to which are added helps at hard places, explanations of seeming discrepancies, and a new system of paragraphs'. What is going on here?

One major feature of the Protestant Reformation was an insistence on everyone's right to have access to God's Word. No longer was the Bible's meaning to be determined by a closed monopoly of Church authorities – ultimately the Pope himself. Translated into the vernacular, it would now be open to every literate Christian to read, and interpret, for herself or himself. This soon unleashed a tide of 'heresy', with, among much else, dramatic political consequences. With the Authorized Version under their belts, English radicals of the mid-seventeenth century turned it into a charter for revolutionary, even utopian, action. Black Americans could place it at the spiritual core of both their endurance of slavery and their liberation struggle. The Bible, especially the King James Authorized Version, becomes a dangerous, ambiguous, inspirational, resource on which dozens of diverse constituencies can draw to create their own 'pure' version of Christianity: Christadelphians, Jehovah's Witnesses, Muggletonians, the Brüderhof, Amish, Mennonites, Congregationalists, Seventh Day Adventists ... and many more. Not a few, of course, migrated to the United States to found their own utopian New Jerusalems. Scofield's move may then be read in one sense as an attempt to recoup this situation, to return authority to the religious professionals, who alone had the knowledge and spiritual insight to understand what the Bible really meant. What Scofield and his colleagues produced was nothing short of a thoroughgoing bid for mainstream fundamentalist Protestant authority over Christian doctrine.[2] Ironically, in order to do this they have to totally tame the faith's unruly and Protean central text. Like shipwrecked Gulliver, it is to be thoroughly immobilized in precisely one posture, held down by webs of cross-references fastened to dogmatic theological stakes. It does, and can, mean only one thing, and far from being 'understood even of the very vulgar' it seemingly required nine learned divines to decide what this meaning was. The underlying paradox that the text is held to be simply literally true *and* demands endless poring over in order to be fully comprehended is never resolved. Nor is the tension between the 'literalist' and symbolic prophetic modes of reading it.

The basic argument that I have presented here is not novel. By chance I recently came across a long-forgotten work by the Revd. E.D. Rendell of Preston, *The Antediluvian History, and Narrative of the Flood, as set forth in the early portions of the Book of Genesis. Critically examined and explained*, which in a fundamental sense makes a similar case from a piously devout perspective. First published in 1850, this was the 1864 2nd edition with an interesting new Preface. I naturally assumed from the title that this was some kind of defence of literal biblical truth against the ever-mounting assaults of geologists and palæontologists, typical of the period, reissued with a new defiant post-Darwin Preface. Not a bit of it. Describing his approach as 'intended to illustrate a new principle of hermeneutics' (p. xvii), Rendell is in effect urging his contemporaries to bite the bullet over the *Genesis* account. As Christians they have to accept the truth of the Bible. But the literal truth of *Genesis* is in conflict with empirical facts. Those engaged in convoluted attempts at reconciling *Genesis* and scientific facts 'appear to forget that those documents were really provided by Divine superintendence; and this having been their origin, they must be accurate in the facts they are intended to reveal' (p. xix). But since this is obviously not the case, 'supposing them to treat of physical occurrences ... we are compelled to seek for the facts referred to in some other phases of created existence'. These are 'in different states of the human mind; its relation to the Creator, and its separation from him' (*ibid.*). In short, *Genesis* contains 'figurative histories of spiritual events, true in their nature and exact in their expression' (*ibid.*). He then proceeds, in hermeneutic spirit, to explain the prevalence of figurative thought in the ancient world. Why Rendell's is now among the 'forgotten' texts I know not, but posterity is no impartial judge, and the reasons for humbler books being forgotten are probably as numerous as the reasons for their being remembered. What is patently clear, however, is that he was quite capable of understanding that the truths that his God intended to convey when He inspired the Bible were of a quite different kind from those that concerned the physical sciences. If a provincial British Anglican minister could grasp this in the mid-nineteenth century, it is a little odd that the insight eludes so many contemporary Christians.

6 Time and space, certainty and incredulity

A cautionary case

Michael Paget Baxter (1834–1910) was in many ways an admirable man: a devout, charitable, Christian and passionate in his advocacy of the fundamentalist creed, to which end he founded a weekly newspaper, the *Christian Herald*. In particular he felt that his divinely endorsed vocation was to interpret the prophetic passages in the Old Testament *Book of Daniel* and, in its entirety, the New Testament *Book of Revelation*. His application and biblical scholarship were unstinting, and he was early on convinced that he had the key. In 1866 came the first edition of his *Forty Coming Wonders of Scripture Prophecy between 1890 and 1901*, a fourth edition appearing in 1880 and a fifteenth, posthumously, as late as 1923 (title wordings vary slightly for different editions). From the 1880 edition onwards this was illustrated with 52 engraved illustrations of the Apocalyptic events that he foresaw – 'Four destroying angels, bound at the Euphrates, loosed for a year, a month and a day' (*Revelation* ix. 14, 15) for example – events which were due to last 391 days from 23 December 1898 to 18 January 1900 and see 200 million demon horses and horsemen slay one third of the human race. There was, however, an obvious problem. The Boer War came and went, the *Titanic* sank, but the bottomless pit remained closed. In the 1923 edition (a reissue of Baxter's last, pre-Great War, version with added biographical material) the engravings and much of the text remain the same, but the dates are all revised, the majority of the wonders now being predicted for the late 1920s and early 1930s: the arrival of the four destroying angels was now pencilled in (with a qualifying 'probably') for 17 December 1926 or 21 November 1928. To be fair, Baxter admitted that his original calculations had been in error, due to an unclear section of biblical text, and he does now provide two alternative dates as reflecting continuing textual ambiguity. His daughter

Violet, renowned for her charity work among London's homeless, who called her the 'Silver Lady', died in early 1973.

Forty Coming Wonders marks the high point of literalist prophetic decoding, unmatched by any of its numerous popular successors for detail, rigour, learning, and subtlety. I once, out of curiosity, delved into it in some depth and discovered that Baxter does not merely deploy a simple 'day for a year' decoding algorithm, but further embeds within this a 'day for a day' algorithm in which the whole plot repeats itself literally during the final days – thus providing him with both an extended historical interpretation of a symbolic kind and a detailed, literal, prophetic one. It would be too easy to mock Baxter's honest and faithful folly, but he was clearly utterly misguided in his labours. It is unfortunate that they are not better known to today's fundamentalists, as they are a cautionary example of the futility of faith, however genuine and intense, when exercised in the service of an erroneous premise. But Baxter's case has further significance, for he surely embodied a certain kind of mentality or psychological character which, as the long-lived popular success of his work demonstrates, was widely shared in the fundamentalist community. The current popularity in the United States of a belief that we are living in the Last Days foretold in *Revelation* suggests that it is still widely prevalent.

Creationist psychology

I believe we need to approach Strong Creationism not as a quasi-scientific theory but as a psychological phenomenon. A number of distinct features of the Creationist mentality can be identified which it will be helpful to explore. It is, of course, a standard rhetorical tactic, used by advocates of many ideological persuasions, patronizingly or contemptuously to pathologize one's opponents. One immediate pay-off is that one feels that it is unnecessary to listen to anything they are saying. I do not wish to be seen as engaging in this kind of game, but there do, on the face of it, appear to be a number of respects in which the psychological character of Strong Creationism is puzzling and anomalous, and these need to be brought out into the open and confronted. (There are also some similar, if less obvious, features in extreme anti-religious materialist thinking that I would describe as 'scientific fundamentalism'. More on these at a later stage.)

Time and space

The first of these features concerns Creationists' views of time and space. They are undoubtedly happy with the notion of 'eternity', for it is eternal joy or damnation which awaits us all after the Day of Judgment, and 'eternality' is one of God's principal attributes. (I leave aside the more sophisticated theological arguments, dating from St. Augustine's *Confessions* in the fifth century, that Eternity refers to existence *outside* time, rather than infinitely extended time. But Woody Allen's quip remains irresistible: 'Eternity is a very long time indeed, especially towards the end'.) When it comes to real time-spans, their attitude is quite different, since they evince a kind of horror, incredulity, or incomprehension of numbers larger than five figures. The cosmic plot is assumed to unravel in a little over 6,000 years (on the Ussher calculation we are already, at the time of writing, up to 6,017). This time-span is quite unusually short among religious cosmologies: the Hindu figure exceeds even the 13.5 billion years since the 'big bang' currently accepted by astrophysicists.[1] A defence of the Creationists' time-scale that would be capable of challenging the current scientific consensus would require a total revision not only of terrestrial geology but of the whole of physics, in order to explain how stars and galaxies appear to be millions of light years distant. It is as if there is a need to constrain what is conceivable by God within the boundaries of one's own human imaginative capacity. Somehow, such vast stretches of time, such 'extravagant duration' as the mid-nineteenth-century writer Sharon Turner put it,[2] seem to strike Creationists as plainly ludicrous, impossible, contrary to common sense. And rather than subjecting this feeling itself to critical examination, they believe that it has sufficient authority to override all evidence to the contrary. To argue simply that because this is what the Bible says, faithful Christians must accept it, is itself unacceptable, for there is no genuine consensus among Christians as to what the Bible should be understood as saying on the matter. Really vast distances evoke similar incredulity from Creationists, who are nonetheless happy with the notion of Infinity.

Historically, as described in Chapter 1, the psychological impact of natural science on Christian understanding of the *Genesis* account and the 6,000-year Fall–Flood–Salvation–Judgment Day plot first manifested itself at a cultural level in the latter part of the eighteenth century and accelerated constantly thereafter. Eternity and Infinity were fine as abstract or quasi-theological concepts, but the new, and constantly expanding, real time-scales and distances being bandied about by 'natural philosophers' created, as it were, a combination of agoraphobia and claustrophobia. The overwhelming scale of the real immensities of space revealed by astronomers threatened to reduce humans to the insignificance of mayflies, hard to reconcile with their cosmic importance in traditional Christian belief, while the apparent absence of any divine agency within this vastness and the rather crude 'clockwork'-like nature of the principles by which cosmic phenomena then appeared to be determined could create a feeling of helpless entrapment. (The widespread receptivity to the comforting Argument from Design was surely due, in part at least, to these anxieties.) In Europe and on the United States East Coast, the majority of educated people gradually learned to live with this over the next two centuries.

In Britain, advocates of Strong Creationism were of course prominent in the 'Darwin wars' of the post-1860 period, but they became increasingly marginalized by the Christian mainstream, who found other strategies for reconciling their beliefs with the rapidly developing scientific cosmology, including its evolutionary component. This is not to say that Creationism ever disappeared in Britain, but its proponents remained on the back foot throughout most of the last century, and by the mid-century a less literalistic 'design' version had largely replaced it.

The immensities of space and time remain awesome, but fascination has now replaced fear and trembling as the standard psychological response, while, in another significant psychological shift, the notion of descent from primates no longer triggers the paroxysms of revulsion and disgust with which it was first greeted (and which some Creationists apparently still experience). Indeed, the higher primates now figure, along with whales, penguins, and meerkats, among everyone's favourite fellow species. But before we become too smug, we ought perhaps to consider how well any of us genuinely grasps the scales of time and space that we so easily refer to. Can even the most sophisticated evolutionary scientist really yet get a proper handle on the way events unfold over millions and millions of years? Yes, we think we have a fairly reliable schematic plot-line, but if we are honest we know that there must have been innumerable convoluted twists and turns along the way, and most probably many contingent factors were involved, of which we as yet have no inkling. Every year brings new surprises which excite and fascinate those in the relevant scientific disciplines. It is only about 50 years since something as basic as continental drift was finally accepted by geologists. Nonetheless, what non-Creationists can now accept with equanimity is that the proposed time-scales and distances themselves more or less correspond to the reality. What is now psychologically puzzling is Strong Creationists' enduring resistance to the notion of time and space measurements extending in years or miles above six figures (and their continued deep revulsion at the thought of animal ancestors).

I feel bound to add a brief coda to this in the light of Christine Garwood's fascinating *Flat Earth. The History of an Infamous Idea* (2007), alluded to earlier: a detailed examination of the strange rise of belief in a flat Earth during the early nineteenth century and its persistence down to the present. Incidentally, it transpires that the common belief in a flat earth at the time of Columbus is a complete myth. Initially promoted by the 'travelling lecturer and quack doctor' Samuel Birley Rowbotham (1816–1884), writing under the pseudonym 'Parallax', it was most fully expounded in his 1865 *Zetetic Astronomy. The Earth not a Globe!* The last serious proponent of the Flat Earth doctrine was California-based Charles Johnson (with his loyal partner Marjory), founder of the International Flat Earth Research Society of America, who died in 2001. While clearly eccentric and marginal, the story greatly illuminates the problematic nature of 'literal' readings of the Bible, for nearly all these 'Zetetic' believers – and none more so than Johnson – shared a deep faith in this approach. They thus concluded, on the basis of various passages in the

Bible referring to Heaven being 'above the Earth', the Earth being fixed on pillars, with water under the Earth, and the Earth being 'outstretched' or having 'corners', that the Bible was teaching that the Earth was flat. From their point of view, anti-evolutionist Creationists who nonetheless accepted the 'globularity' theory were guilty of evasion, baulking at following their convictions through to their logical conclusion. Psychologically it is very interesting to note how much of the 'Flat Earth' rhetoric plays with the 'insanity' of believing the Earth to be a globe hurtling through space at thousands of miles an hour. There is also an apparent inability to grasp that 'up' and 'down' are not objective cosmic dimensions but simply rooted in the physiological mechanisms which serve our sensations of physical orientation. For Johnson, one knock-down argument was that his Australian-born wife 'knew' she had not been upside down in the land of her birth! As for rewriting the total scientific cosmology, he was eagerly up for this, claiming the entire Copernicus–Galileo–Newton–Darwin–Einstein succession to be a superstitious plot to addle the human race.

Garwood notes that, when the Flat Earth consequences of biblical literalism have been pointed out to them, Creationists have reacted to this as a slanderous insult. But the Flat Earth case only highlights the fact that totally thoroughgoing literalism is impossible, and that there is a continuum between 'literal' and 'symbolic' modes of interpretation. This being so, Creationists can surely be legitimately required to explain the criteria that they use for drawing the boundary line between the two kinds of passage, and why the Genesis account falls on one side but Flat Earth passages fall on the other. Geneticist Steve Jones's puff for Garwood's book urges 'Every creationist should read it', and I can only endorse this sentiment.[3]

The social psychology of Creationism

Where then has the resurgence of Creationism come from, and why? A hypothesis of a social psychological character suggests itself. Few would dispute that the heartlands of contemporary Christian Creationism are the Deep South and Mid-West of the United States, with additional support coming from evangelical churches popular among ethnic minorities. In Chapter 1 we noted that in the USA it had always remained powerful in these regions, as the famous Scopes trial in Tennessee and the popular Creationist geology texts of George McReady Price bore witness, while the inclusion of evolution in school biology curricula continues to be controversial in some states and counties. To elaborate a little on the points made previously, my perception of this is that historically the populations of these regions kept themselves relatively insulated from the main tides of European scientific and intellectual culture until the late twentieth century. Their roots frequently lay in communities founded by evangelical, usually Protestant, religious exiles, dissenters from the dominant churches of their European countries of origin. Their rural life-styles were arduous and often insecure, with religious faith providing their primary source of psychological strength. This was the country of the itinerant fire-and-brimstone preacher, of Born-Again Christianity, and periodic open-air revivalist rallies, which one suspects offered a welcome break from the tedious and laborious routines of everyday rural life. Dozens of largely autonomous Baptist, Seventh Day Adventist, and Pentecostal churches and chapels flourished – and continue to do so. To take at random the small town of Shelbyville, Tennessee (population just over 16,000): among the churches listed on its website are such names as Hilltop Church of the Nazarene, Life Worth Living Baptist, Whittaker Church of Christ, ten further Baptist churches, and the rather cryptic Bright Temple Cogic (*sic*) and Raus Church of Christ.

Evangelical American fundamentalism was not initially centred in these regions, however. It was in the northern states that the impact of modernism was first strongly felt in the latter years of the nineteenth century, and from the 1880s to the early 1920s it was cities such as New York, Boston, and Chicago which witnessed a campaigning resurgence of fundamentalism and anti-evolutionism. The mainstream churches were riven by fierce and acrimonious contests between their more liberal establishment leaders and the newly vocal fundamentalists. By 1920 a more widespread reaction (not all of it fundamentalist) had set in against Progressivist liberal Protestantism, which was regarded as too anaemic and intellectualized. The fundamentalist argument centred not only on scientific issues such as evolution but, perhaps even more heatedly, on the status of the Bible itself. While liberal believers felt able to come to terms with the 'Higher Criticism', accepting the Bible's multiple authorship and the non-literal character of much of its earliest content, for the fundamentalists this was sinister heresy and betrayal. As indicated in Chapter 1, this first fundamentalist wave had more or less lost its momentum by the late 1920s. But while traditionally fundamentalist in their understanding of the Bible, many of the rural west, mid-west, and southern communities long remained relatively insulated from 'modernism' beyond the improvements that it had brought to agricultural machinery, fire-arms, and the advent of the automobile.

Such communities have always seen themselves as simple, honest, devout, hard-working, and self-reliant (which they mostly are), but this source of psychological strength came at a psychological price. It was easily accompanied by anti-intellectualism, morally infused scorn for big-city culture, and isolation from events beyond their own restricted locality. These attitudes readily facilitated a survival of the earlier Creationist mentality and, given the shortage of printed books, engendered an extreme reliance on the Bible as the sole, unchallengeable, authority on all things, a reliance which had long withered across Europe (including Catholic regions) and the urbane North-East from Maine down to Washington. The old six-millennium time-frame never met any serious local challenge, and if it had become doubted by those in the sinful cities of Europe (from whence many of these believers' ancestors had been driven) and the Eastern states, this but reinforced a sense of belonging to the handful of genuinely faithful elect.

But things could not stand still for such communities. Even as pride in their own fundamentalist religious culture and values became ever more intense and conservative, the technological forces at large in the outer world increasingly penetrated theirs. With the US rise to global pre-eminence after 1945, and technology's astounding advances pouring forth in accelerating abundance, the involvement in national affairs of these proud, self-confident, ardently religious but also, as a result of long marginalization, resentful communities became inescapable. As 'modernism' finally hit with its full force, it was perhaps even more traumatic for them than it had previously been for devout northerners. There was a dramatic rise in immigration from the northern states, as industrialization and urbanization began to take hold. Expansion of higher education meant that their young people were also encountering the scientific culture in all its optimistic mid-twentieth-century glory. The new mass media meanwhile provided Creationist evangelicals with an opportunity that they were eager to grasp as a means of holding the line against materialist heresy. The USA has always been a deeply religious society, albeit highly variegated, and the majority of its universities and colleges are also religious in foundation. For many decades this meant that even in its intellectual and scientific centres of excellence a religious resistance to aspects of modern science remained common, even if not Creationist in character. And most non-religious academics also perhaps felt obliged to treat the ardently religious with more respect and courtesy than they have often done in Europe. Karen Armstrong (2000, Chapter 3) offers a very insightful analysis of the extent to which US culture has, during and since its foundation, been profoundly divided between secularist and religious understandings and interpretations of its character and destiny. Addressing this now would be a detour too far, but it is essential reading for those wishing to pursue in greater depth the matters raised here.

The present situation therefore appears to be somewhat as follows. At a time of global environmental and political crisis with which science is unable to cope, and when scientific knowledge itself has become extremely arcane in character, a large constituency of Americans from the background described above find themselves having to engage with, and embroil themselves in, the full scientific cosmology more or less for the first time. Largely ignorant, it has to be said, of how the same encounter was initially played out over a century ago in Europe, and ignorant also of the history of Creationist thought itself, as well as knowing little of twentieth-century theologians such as Rudolf Otto and Paul Tillich, they are once more circling the wagons and reacting against their beleaguerment with every weapon to hand (including many that their enemies invented by accepting the very scientific cosmology that they oppose). In doing this they have natural allies among those whose previous historical experience of social marginalization and reliance on a simple form of evangelical belief has been very similar, and often of even greater intensity. This is especially evident, both in the USA and in the popularity of Creationist anti-evolutionism in British evangelical African–Caribbean churches.

Sociologists and historians of religion have long observed and understood how evangelical and fundamentalist beliefs typically originate and prosper among the socially oppressed and marginalized. Saying this does not, of itself, mean that such beliefs are wrong or inferior. But a problem certainly arises when, as now, these beliefs *do* include a central doctrine which is demonstrably erroneous, and its proponents are no longer marginalized but have become an integral component of mainstream social life. This difficulty cannot, however, be overcome by contempt, patronization, pathologization, and reciprocal enmity, although it is hard even to articulate it without the risk of appearing guilty of one or other of these sins. None of us is responsible for where we find ourselves initially located culturally, geographically, socially, economically, or ethnically. And this will quite naturally determine those beliefs, attitudes, and values, the 'world view', in short, from which we start. But the more psychologically central certain elements of that world-view are, and the more crucial for our sense of identity and self, the harder they will be to relinquish when we meet challenging alternatives, and the more tenaciously we will strive to cling to them and find ways of reconciling them with, or rejecting, such alternatives.

Few things are more fundamental, albeit elusive, to our world-view than our senses of space and time. The 'Strong Creationist' position, reliant on the calculations of two English divines in the early seventeenth century, entails views of space and time which are quite simply as wrong as the idea that the Sun orbits a static Earth. To succeed in establishing the literal truth of these views would involve rejecting virtually everything that science has claimed to discover over at least the last three centuries, and that includes the knowledge which enables us to make televisions, mobile phones, and MRI-scanners.

The attitude towards the Bible which underpins Strong Creationism is, to be frank, a superstitious one. We are all to a greater or lesser degree superstitious, but it behoves us to admit the fact and strive to combat our superstitions when we spot them. The very plethora of non-Creationist versions of Christianity should reassure Creationists that admitting and abandoning theirs need not amount to a leap into utter atheistic apostasy.

Certainty, incredulity, and doubt

But Creationists *know* they are right. Knowing that one is right does not necessarily mean that one is, of course. People in all ages and of all faiths have 'known' they were right about things on which we, for our part, 'know' that their 'knowledge' was mistaken. The nature of 'knowing' is a complex philosophical issue into which I have no wish to enter here at any length, but a few points do have to be made. The first is that we use the concept in a variety of different ways. At one extreme the term is actually quite redundant, as when we say 'I know I am sitting at my PC', or 'I know I exist' – in normal circumstances the term 'know' adds virtually nothing. 'I am sitting at my PC' and 'I exist' are quite sufficient (in these cases it is the 'I *don't* know' variants which are significant!). Another usage, probably the most common, refers to bodies of facts or techniques that one has acquired: 'I know the 12-times table', 'I know Paris is the capital of France', 'I know how to play the piano', 'I know the way to Amarillo'. If your teachers were in error and you never had cause to doubt them, this knowledge might sometimes be wrong, but its rectification would rarely involve rethinking your entire world-view, moral values, or sense of identity.

But there is a third kind of usage in which, when we say 'I know', there is a kind of added insistence on its intensity or psychological importance. Rather than being based on ascertainable evidence, the knowledge in question is intuitive, based on a feeling or 'sixth sense': 'I *know* he's being unfaithful', 'I *know* I'm doing the right thing', 'Even before I arrived, I just *knew* something was wrong'. And there is a yet further context in which the use of the term can be problematic: this is when we are referring to our own subjective experiences. At first sight one might think that in these circumstances it is impossible to be wrong: these are our own, direct, experiences, after all. I cannot be wrong when I say 'I know I am in pain' (although again this hardly says more than 'I am in pain', and the philosopher Ludwig Wittgenstein used this as an example of a misleading use of the term). But what of a widower who says, of his dead wife, 'I know I heard her speak to me last night'? These are deep philosophical waters, but it may be helpful to note that two things are going on here: the 'raw' experience, as it were, and what we say or believe that experience was or means. And it is a moot point whether we can ever entirely 'know' the first in its pure form. At any rate, in order to communicate about it we have to use the language of our culture and time. This is actually to return to the earlier point about there being no entirely theory-free 'facts', for this language always carries with it implicit quasi-theoretical assumptions about the nature and meaning of the subjective experiences to which it refers.

The feeling of certainty which can characterize religious beliefs is often based on having had some subjective 'religious' experience, as a result of which one feels entitled to say, for instance, 'I know that my Redeemer liveth'. That these experiences are profound and meaningful I do not dispute, nor am I inclined to 'reduce' them to simple 'mislabelling' of the sensations accompanying physiological events. What I do want to insist on is that if we are not careful we can easily read far more into their bearing on specific religious beliefs and tenets than they really warrant. This is especially easy if we have no knowledge (of the second kind identified earlier) of comparative religion and history of religion. When we do have this knowledge, we are in a much better position to evaluate the experience and the limits of its doctrinal implications.

I suggest that the Creationist, whose over-riding frame of reference is the Bible, the teachings of which have provided his or her life with depth and meaning in a whole range of ways, has fallen into the trap of treating the Bible as therefore literally true in *all* respects. In other words, he or she has concluded from subjective experience of the Bible's value as a major source of comfort and guidance that this confirms its divine character in its entirety.

This, as we mentioned earlier in the case of Henry Morris, is coupled with a belief in the value of certainty, reflecting a deep psychological need for some infallible, indisputable, core knowledge. Not everyone shares this need, but those who do perhaps find this inconceivable. 'You must believe in *something*!' they might exclaim. Lacking such a central fixed point, a person must, in their view, be adrift and helpless.

In the context of the Creationism vs. Evolution debate, or the broader Creationism vs. Science debate, this has a very serious consequence. The two sides are talking at cross-purposes. The Creationists' premise is that they are absolutely right; what they 'know' has been revealed by God and is not up for negotiation. Their debating task is thus to reassure both themselves and their supporters by ritually proving their immunity to doubt, and finding some argument, however convoluted and implausible, that appears to show that the opponent's case still has a theoretical loophole or can be destroyed by a single apparently anomalous 'fact'. One is facing a spiritual enemy, a witting or unwitting agent of Satan, over whom crushing victory is really the only acceptable final outcome. The scientists' premise, by contrast, is that actually all knowledge about the physical universe is, in principle at least, provisional. What they are looking for is evidence of there being grounds for genuine doubt about the adequacy of their current theories. I do not wish to idealize scientists as saintly, ever-open-minded heroes. In reality they can be as stubborn as anyone else, fight tooth and nail to defend their position, and even play dirty on occasion. And this is probably truer now than ever. Even so, if the weight of evidence is strong enough, scientists have to yield and admit that they were wrong. If they do not, they risk marginalization by their peers and even effective eviction from the scientific community. But in making this admission, while possibly deeply disappointed, their belief in science itself is not challenged. It is, or should be, more like losing a chess game than losing one's faith.

In short, Creationists are for certainty, scientists (in their own eyes at least) for doubt. Or rather, the only scientific certainties are negative ones: the recognition that something is certainly *not* true. For Creationists, doubts are signs of weakness; for scientists they indicate where the next research action has to be focused, and as such they are positively valued. On one side we have a group for whom it is inconceivable that a life not lived within and guided by a religious faith can be meaningful or even moral, on the other a group for whom this very need is a kind of immature, superstitious, and irrational infantile dependency, at least when it comes to understanding the physical universe. Neither group actually represents the Christian or scientific communities in their entirety, and the more extreme and vociferous pro-evolutionists can appear just as 'fundamentalist' as their opponents, replaying late Victorian debates but unfortunately knowing far less about religion than T.H. Huxley and his fellows did. And ironically, as we saw earlier, in their insistence on the Bible's factual literality, the Creationists have unwittingly bought into the thoroughly materialist notion that only factual texts are meaningful or valuable. The difference is that for Creationists the Bible is true in an absolute and immutable fashion, whereas for scientists their 'truths' are always, in principle at least, incomplete or only provisional.

Each side is genuinely puzzled by the other, but they cannot find a route for resolving the puzzle in the face of each other's stubborn intransigence and, as they see it, irrationality. This has a more serious consequence, for it has the common psychological effect in such circumstances of each then projecting on to the other their own, in Jungian terms, 'shadow'. For Creationists the evolutionists become servants of demonic materialist atheism; for scientific evolutionists, Creationists become the embodiment of bigotry, ignorance, and superstition.

To reiterate what was said earlier, accounts of the nature of the physical universe are intrinsically transitory; they have a built-in obsolescence. Every 'closure' (like the final discovery of the structure of DNA) simply opens a new set of doors. Enduring texts are not, as we saw in Chapter 5, of this kind at all. Greek drama can still be performed with enormous impact, Homer's *Iliad* and *Odyssey* can still inspire films and new translations, but only historians of science are interested in Aristotle's views on botany and biology or Ptolemy's astronomy. Ancient mythologies and creation stories, be they Norse, Egyptian, Babylonian, or Hindu, remain interesting in a way in which Dalton's chemistry, Harvey's writings on the heart, and Pasteur's research on the nature of disease can never be, except to historians of science. Only occasionally (Darwin, interestingly, being a case in point) do past scientific writings remain widely readable and fascinating, and that is not because of their content but because of their literary quality, the attractiveness of the authorial voice, the author's personality and attitude to his or her labours. The Bible's enduring richness lies in just this direction – as an archive of stories, poetry, moral insight, and accounts of enigmatic events. Were it all like the passages from *Numbers* cited earlier, it would be as dreary as the endless records of wheat and cattle traders that comprise so many of the cuneiform tablets from Sumer and Babylon.

Faith

The term 'faith' always carries the connotation that it is a belief or trust that goes beyond the evidence. Like the emphatic sense of 'know' mentioned above, it carries the implication of being a little beyond reason or beyond the evidence. This does not mean that it is irrational, but rather that it is non-rational. Knowledge and faith of this kind are absolutely essential for human life: faith in our friends and partners, faith in our leaders, faith in the honesty of those we are dealing with, faith in the justice system or in doctors, and, in the case of science, faith that one's research methods will yield the answers that are sought. But in all these circumstances our faith can be broken, events compelling us to recognize that it was misplaced or mistaken. Religious faith is not of this kind. It is unconditional and absolute, admitting of no refutation or falsification. When people lose their religious faith, it is because it simply no longer works for them psychologically; it is not directly entailed by worldly events and experiences, even though it may typically be triggered by them. Catastrophic wars and disasters may lead to many losing their faith, but there are always others whose faith survives or is even strengthened. Some actually 'find' their faith in such circumstances.

There is no refuting the validity of a religious faith (as opposed to specific factual claims) in a 'rational' scientific fashion. Were it so refutable it would, by definition, not be a faith in the full religious sense at all, but a kind of scientific theory or hypothesis. But equally, of course, there is no way in which a religious faith can be scientifically confirmed and properly remain a faith. Which returns us to the point on which I closed the previous chapter. As Southern Catholic author Mary Flannery O'Connor allegedly said, 'Faith is that which you know to be true, whether you believe it or not'.[4] None of this implies that reason and faith are entirely unrelated. Faith is not, though, *produced* by reason; rather reason is constantly brought to bear upon it, reshaping religious doctrines in the light of new ideas, historical experiences, and knowledge. It is also used to provide justifications for these doctrines. These tasks are, basically, what Theology is about.

In some circumstances, or for some people, reason may indeed succeed in demolishing a faith, but it is commonly remarked that nobody was ever converted to a faith by reason alone. Faith, in short, always precedes whatever contributions reason may make to its specific form. Most poignantly, one of Darwin's leading followers, George Romanes (1848–1894), underwent tortuous and convoluted spiritual struggles before rationally deciding that his inflated reason was the barrier to religious belief, finally re-converting (according to some accounts) to Christianity on the eve of his premature death. (Another indication of their differing logical status is what linguistic philosophers would call the 'grammar' of the two terms. For instance, we commonly talk about having faith in something, but never about having reason in it. And losing one's faith is quite different from losing one's reason!)

Complexity phobia

One further psychological point needs to be made. The world is bewildering and complicated, and navigating our way through it is no easy task. We are faced with difficulties from all directions. Finding a sense of our own self-identity and role within the world is not the least of these, and it ramifies out into our social relationships, economic circumstances, historical events, and a constant plethora of often contradictory incoming information about everything from politics to health, show-business to religion, our employer's intentions to global warming, crime on the street to disasters in far-off lands. A handful of individuals relish this mayhem, and most eventually learn to manage it more or less effectively. But for some it is all too much. The specific life-circumstances of particular individuals and groups render it, quite genuinely, psychologically impossible for them to take on board the complexity of the whole and the demands which doing so would make. Perhaps none of us now knows quite how to handle it (certainly not me).

One response to this is what I would call 'complexity phobia', leading to a mode of life in which everything is simplified. This may take the form of (for example) addiction to heroin or crack cocaine: however miserable and desperate the addict's life is, it has the compensation of being simplified down to a single focus. Another form, I suggest, is religious fundamentalism. This is not as extreme as drug addiction perhaps, and the analogy is maybe a little cruel, but it can serve a similar simplifying function. Issues of self-esteem are resolved at a stroke, while the intellect need no longer try to encompass the evident complexities and uncertainties of the ways in which our physical and social worlds operate. To an outsider it appears that the intellect becomes focused in vestigial fashion on obsessive circular pondering on a single religious text. Moreover, the often intolerable demands on our compassion and sympathy which the modern world makes, and the 'compassion fatigue' that most of us sometimes guiltily feel, are lifted. What happens is God's will. This can have quite opposing effects. It may enhance the ability to engage in responding lovingly to the world's suffering without becoming depressed by a feeling of helplessness. But in some instances it may promote indifference, a belief that God in any case cares only for the 'elect', of which one is now a blessed member. And everyday decision-making is rendered easier by rigid rules of conduct. This is surely part of the appeal which some aspects of Sharia law hold for the wretched of the Islamic world. In sum, the horizon narrows and life can be lived in the service of a single unproblematic cause. There is then a sense in which Creationists are actually opposed to increasing knowledge. They only want to know that those who try to do so are wrong and that their own current body of knowledge requires no expansion.

Fundamentalist religious belief is not the only form in which these simplification tactics manifest themselves, but it is a particularly clear example of them in operation. The Creationist/ID conviction that complexity of itself signifies super-ordinate design reinforces the argument. It lets its proponents off the hook of having to face the consequences of admitting that it usually signifies quite the opposite, as was pointed out in Chapter 2. No doubt we all engage in such simplification tactics to some degree, but not to such a self-imprisoning extent.

What all this is pointing to, then, is again that we are dealing with a psychological issue. The extreme Creationist and more passionate materialist positions are each attempts to prove the complete sufficiency for all human needs of one pole of a bipolar axis: of, in Karen Armstrong's terms, '*mythos*' against '*logos*', or the other way round. By seeking to demonstrate how all the other's concerns can be assimilated into their own framework, they are engaging in an exercise akin to hovering over the South Pole in a balloon and declaring that really everything is in the North. Instead of seeking ways of integrating the opposing modes of psychological functioning which they represent, they simply cast the other as a deluded and dangerous enemy. In psychological terms it must be said that this is potentially very dangerous indeed. It is not one camp or the other that we should 'pathologize': rather the very debate itself as currently conducted is symptomatic of a deeper psychopathology. What this may be I will touch on in the final chapter.

7 Psychology and the real enigmas

I have tried in previous chapters to illustrate that many physical scientists are at cross-purposes with radical Creationists or ID exponents. One group seeks an ever-more detailed and elaborated account of the workings of the physical universe, the other a static unchallengeable framework of cosmic meaning for all human life. It is not the job of the physical sciences to provide a philosophy of life, even though many scientists have thought that it was; nor is it the job of religion to provide anything else, although again many religious thinkers have thought that it was. This is not to deny that the motives underlying the scientist's choice of vocation may be rooted in a broader 'philosophy of life', which could indeed be a religious faith. Such motives will play a part in determining their research agendas, but will be absent from their scientific methodology as such. Christian, Marxist, Muslim, and atheist chemists alike accept the same 'laws of chemistry' and utilize the same research methods and technologies. Conversely a satisfactory 'philosophy of life' should be consistent with the findings of science, although, unless it is an attempt to erect a philosophy on a purely scientific basis,[1] it cannot accept them as totally sufficient for its own purposes. To put it very crudely indeed, physical science is about matter; philosophies of life, including religion, are about mind (including meanings and values as mental phenomena).[2]

There is, however, one ostensibly scientific discipline which does purport to address mental phenomena, and that is psychology. The precise status and nature of this discipline are a matter of perennial debate, and it is so diverse in the topics that it studies that it is really little more than an umbrella label. But, for complex historical reasons, until around 1990 one issue that it largely either evaded or dismissed as a non-scientific philosophical question was, surprisingly, the relationship between mind and matter, or the 'Mind–Body Problem' (MBP) as it is familiarly known. While psychologists have, of course, researched the mental consequences and correlates of physiological processes, the actual nature of 'the mind' itself has invariably been left to one side as being the business of philosophy. Usually the phrase 'Mind–Body Problem' is applied in the narrower sense of how our individual minds are related to our organic bodies, but Western philosophers and theologians have long felt there to be a broader riddle: the place (if it has one) of mind in nature. Whether broadly or narrowly addressed, the nature of 'mind' remains a genuine enigma. It is often equated with 'consciousness', but most would now accept the notion of an unconscious level of the mind, so the two are not synonymous – which only exacerbates the difficulties of getting to grips with the issue.

This is not the context in which to review the various difficulties and paradoxes besetting attempted solutions of the MBP, but it is pertinent to stress its curious formal resemblance to the 'Religion vs. Science' issue – which we might perhaps term the 'God–Matter Problem', or GMP. In both cases, as I see it, physical science finds itself unable to gain a purchase on the issue in a productive or satisfying fashion, precisely because the topic is, almost by definition, one which physical science is not designed to address – one might even say, is positively designed to ignore. The major difference is that while the existence of mind, in some sense at least, is indisputable, that of God is not, so there is a question mark over the very legitimacy of the GMP. The GMP is not necessarily equivalent to the issue of the 'Mind's Place in Nature', however – although inevitably any view that there is either a transcendent or an immanent presence of mind in nature almost inevitably ends up endowing it with a divine character.

The physical sciences have perhaps rather lost sight of these issues during their sensational successes over the last two centuries. Consequently they have felt driven to account for both mind and religion in simple reductionist terms. The advent of the study of psychology towards the end of the nineteenth century nevertheless did rather complicate things. Now we had a purported science explicitly concerned with the mind, completing the 'circle of the sciences' (a common expression at the time). Since then psychology has, as already noted, developed into a patchwork quilt of a discipline, its very variety reflecting that of its subject matter. It has, along the way, yielded a wide range of pragmatically useful insights and falsified many myths, but it has utterly failed to achieve a consensus at a theoretical level. While it is widely held that psychology's status as a science is anomalous, this is not necessarily to condemn it: indeed, it is reasonable to argue that it is *necessarily* anomalous. The author's own position may be gauged from G. Richards (2010),[3] and a detour into this complex issue is out of place in the present work.

Psychology is, even so, relevant to us because it marks the meeting point between the two kinds of concern referred to earlier. It must, we feel, be about the meanings of human life *and* about how to treat these 'scientifically' (hence, among other things, its intimate involvements with psychiatry and psychotherapy). It is a tall order. But one consequence is that, even if some psychologists do not wish to, psychology has to take mind seriously. When it has refused to do so, as in the case of the early twentieth- century school known as 'behaviourism', which famously strove to eliminate 'mind' from consideration altogether on the grounds of its inaccessibility to objective scientific study, it rapidly loses its credibility. The reductionism which has characterized so much physical science has then been a perennial bone of contention within psychology, and many associated with the discipline, including the present author, believe that the role of reductionism in psychology is a limited one. This doubt includes Evolutionary Psychology. To quote Roger Smith (2007):[4]

> In so far as the evolutionary story by itself is supposed to tell us what human nature really is, it fails, because the story is so abstract and detaches itself from the historical story about what has made the particulars of what people are. Whatever human inheritance is universally shared, it has no expression outside human development. This is a cognitive matter; but it is also a moral one, since to value a person is to value a particular person not the abstract entity, humankind. And we value a person, including ourselves, by placing her or him within a significant story which is about that person's life and surroundings, not by identifying the person as an evolved animal undifferentiated from others who share the same remote ancestors. (pp. 241–2).

Moreover, 'final causes' in the form of conscious purposes and intentions are absolutely integral to the way in which responsible humans understand, and account for, themselves and others. A reductionist psycho-neurological or evolutionary explanation usually becomes relevant only when something goes wrong.

More to the point, contrary to what might be assumed, relations between psychology and religion (especially Christianity) have, historically, rarely been fully confrontational. I will say a little more about this later, but might remark here that the past 20 years have seen the reappearance in much psychotherapeutic discourse of words like 'spiritual' and 'sin', and the task of counselling and psychotherapy has often been cast as aiding some sort of 'spiritual quest' or 'spiritual healing'.

That aside, when we do take the mind question seriously, a number of real enigmas soon force themselves on our attention, and these in turn potentially bring religious issues back into the picture. The first group concerns the *irrelevance* of physical reductionist explanations in the case of a number of well-attested, often very profoundly meaningful, psychological phenomena which raise questions that are more of a religious character than a physiological one. Of course, physiology and physics are always somehow involved, but that is beside the point. You do not usually explain why someone is running to catch a train by referring to their adrenalin level: it is simply not that kind of puzzle. But the phenomena in question are 'enigmatic' in character, not simply because we cannot explain them 'scientifically' in terms of physics and physiology (indeed, we can often go some way towards doing so), but because of the inherent quality of a kind of direct and immediate meaningfulness that they are experienced as possessing.

'The Unseen World'

A bit of history. The expression 'The Unseen World' became very popular from the late nineteenth century down to the 1920s. It neatly captured the then popular idea that somehow there is another 'plane' (a popular word at the time) of existence alongside, and normally independent of, our own, usually characterized as 'spiritual' in nature. It was not *quite* independent: under certain circumstances, such as in séances or when loved ones died, the two worlds could intersect or interact. The mediating factor between the two was ultimately mind. It was to this Unseen World that our minds passed after death, it was via the minds of mediums that the departed could communicate with the living, and it was the Unseen World that enabled telepathic communication and clairvoyance.

Initially the notion of the Unseen World had a good deal going for it. In particular it was scientifically plausible in an era when physicists were regularly discovering hitherto invisible forms of energy such as electromagnetism, radio waves, and X-rays. Some of these scientists, such as Lord Stokes and Oliver Lodge, themselves became closely involved with spiritualism and psychic research. Secondly, this plausibility supplied a credible replacement for the earlier Christian notion of a divine, transcendent, spiritual realm which had once been physically located in the celestial realm (or, in the case of its Satanic opposite, under the ground) – a notion which astronomy had rendered untenable in the early seventeenth century. (The attitude of Christians towards Spiritualism was extremely mixed, ranging from keen support and active involvement to horrified condemnation.) Thirdly, it provided an explanation for apparently 'paranormal' or 'psychic' phenomena otherwise beyond scientific treatment by virtue of their capricious unpredictability.[5]

The phrase has long fallen into disuse, but some of the phenomena addressed by 'Psychic Research' (a broader project than Spiritualism and not necessarily committed to its validity) remain genuinely enigmatic, continuing to resist any facile materialistic reductionist explanation. These are of various kinds. Most dramatic perhaps are what may be called cases of 'displaced consciousness': 'near death' and 'out of the body' experiences (often referred to as NDEs and OBEs respectively). Then there is a variety of 'knowledge at a distance' phenomena such as telepathy, clairvoyance, and precognition. Ghosts, poltergeists, and other happenings associated with 'hauntings' are another category. There are also the kinds of meaningful synchronicity phenomena mentioned earlier and, to bring things closer to orthodox religion, the reported beneficial effects of prayer on the sick, sometimes at a distance. Some would want to include 'possession' phenomena (which might include mediumistic trances and events in séances). Finally there are religious mystical experiences themselves, seemingly direct encounters with the divine.

What are we to make of all these? Certainly a high proportion of reported cases in most of these categories can be fairly fully explained in reductionist terms as due to neurological conditions, unusual environmental factors, or simple ignorance and mistakes on the part of the person reporting them (although, as ever, they may present other kinds of puzzle). In recent years a great deal of research has been undertaken on 'hauntings', for example, which suggests that physical factors such as electrical fields, low-level vibrations, and lighting conditions are involved. How far these can explain the actual content of haunting phenomena when they go beyond vague feelings of 'a presence' or intense spookiness is less clear. On other claims the scientific jury is clearly still 'out': NDEs, for example, of which more cases are now reported than in the past, due to advanced surgical techniques in which patients can be recovered from clinical death. These are especially fascinating because of the common forms which they take (a white light at the end of a tunnel, or encountering dead family and friends), and their correspondence to some of the events in ancient myths. Quite how a memory for these can be laid down in a non-functioning brain is in itself mighty puzzling. Even as staunch an unbelieving materialist as the philosopher A.J. Ayer felt his intellectual pride and confidence severely shaken by such an event. Several physical hypotheses have been proposed to account for NDEs and OBEs, but they are currently purely speculative and in any case do not address their sheer profundity of meaning (especially in the case of NDEs) for those who have experienced them.

There are vast literatures on all these experiences, so I will not review them here. The important features that they share are that they are (i) unpredictable; (ii) often extremely meaningful, even life-changing; (iii) often associated with situations of psychological crisis (although this varies considerably); and (iv) currently either inexplicable, either partly or in their entirety, by physical science, or puzzling in ways which physical science cannot address, and which mainstream psychology cannot adequately explain either. While the balance between naïve gullibility and dogmatic scepticism is difficult to strike, I feel that the weight of evidence, admittedly by its nature mostly anecdotal, is that such phenomena cannot *all* simply be dismissed as fraudulent, or delusory, or errors in perception.

Nor would such a dismissal be acceptable to the thousands of people (perhaps the majority?) who have ever suddenly 'known something was wrong' with a loved one, had a premonition of a forthcoming event, or felt comfort emanating from a seemingly profound inner source. It would be wiser to explore their implications for the nature of whatever it is that we refer to by the term 'mind'.

The difficulty facing psychologists and neurologists is to provide orthodox scientific explanations for these phenomena without at the same time 'reducing' them or demeaning their significance. H.G. Wells's excellent short story 'The Country of the Blind' (1904) is probably now unknown to many readers, but at one time it could be 'taken as read' by nearly all literate people. It remains a powerful cautionary tale and is apposite here. The sighted hero finds himself trapped in a society where everyone is congenitally blind. His low sensitivity to sound and his strange accounts of what the world is like lead to his diagnosis as suffering from a disorder caused by his strange eyeballs, and the proposition to cure him by removing them. (He does manage to escape.) That this may be an accurate analogy for reductionist physiological explanations of profound 'religious', 'spiritual', or 'mystical' experiences needs to be kept in mind. After all, they were right: the hero's behaviour was indeed due to his eyes.

All I will say here, without intending to commit myself in any way on their 'validity' or implying that they are supernatural or 'paranormal', is that these phenomena suggest that our current understanding of the nature of 'mind' leaves much to be desired, and that the methods of the physical sciences (largely adopted and adapted by mainstream psychology) are, precisely because they have evolved to tackle the physical world, inappropriate for improving it. They are, even so, insufficient in themselves to justify any sort of leap into specific religious beliefs. But what of the broader question of 'mind's place in nature'?

Mind in Nature

Allusion was made in Chapter 1 to the way in which, as the Paleyite 'Argument from Design' lost its power, many religious thinkers shifted to a position known as Theism. This held that belief in God's presence in Nature was necessary, not because natural phenomena were so stunningly engineered, but on more complex logical grounds. The argument, as propounded by the eminent Unitarian James Martineau, one of its leading exponents, ran roughly as follows.

The natural world operates according to principles of causation and 'force', describable by natural laws. 'Cause' and 'force' are, however, abstract concepts produced by the human mind, rooted in its own experience of being an agent or agency. It therefore becomes unacceptably paradoxical to hold that while mind is necessary for the creation of these very concepts, it is absent in the physical universe, which is only explicable by invoking them. As Martineau pithily observed, 'If it takes *mind* to construe the world, how can it take the negation of mind to constitute it?' Later writers have pointed out some basic flaws in Martineau's own argument, but the underlying sense of paradox has not entirely dissipated.[6] If humans are capable of conceiving of the vastnesses and minutenesses of nature and all its mechanisms, how can mind, the principle facilitating this capacity to conceptualize and understand it, be entirely absent from nature itself? Or confined to the one rather physically and cosmically insignificant life form, *Homo sapiens*?

Admittedly the jump from mind to God is rather a large one, but the notion that mind has some mode of universal presence, and the sense that denying this generates some sort of paradox or contradiction, has proved very difficult to dislodge. This feeling is perhaps reinforced when we realize that virtually all the mental or psychological concepts by which we know and experience ourselves originate by applying to ourselves the concepts and language used to refer to the events, objects, and physical properties that we find in the external physical and social worlds.[7]

Precisely what role the presence of mind in nature might play is another matter entirely, and certainly physical science finds no job for it to do as far as physical phenomena are concerned. Even its proponents are hard pressed to be specific; if it is an initiating agent, then its work is done as soon as the show is on the road; if it is some kind of 'sustaining' agent, then one is led into the kind of Idealism proposed by Bishop Berkeley, who held that, since in order to exist something had to be perceived (i.e.

experienced 'in the mind'), the universe as a whole must exist in the mind of God. 'Supernaturalism', the claim that God occasionally 'miraculously' over-rides 'natural laws' either to aid believers in times of crisis or simply to demonstrate His existence, has long fallen from favour among Protestant (though not Catholic) theologians, having once been a core Christian doctrine. In many respects, dumping supernaturalism was necessary for Christianity's mid-nineteenth-century intellectual survival at a time when sceptical science had become the major cultural phenomenon of the age.

The question of 'The Mind and Its Place in Nature', as C.D. Broad called his comprehensive 1925 review of the MBP, continues to be enigmatic, and again one is inclined to suggest that it is our inadequate concept of 'mind' that is the stumbling block. My own suspicion is that our error is in treating mind itself as an entity or empirical phenomenon, and that there is a sense in which Berkeley was on the right track. Why not consider mind itself as a dimension, along with space and time? After all, it shares much in common with them: it is virtually devoid of empirical properties, it lacks energy, and it is a place where the universe happens.[8] Since the cosmologists now tell us that 11 (or is it 13?) dimensions got squeezed into four in the moments after the Big Bang, there are at least half a dozen or so going spare! Where this would leave the Mind-as-God (or God-as-Mind) notion advanced by Theism is something else, but 'Mind's Place in Nature' remains a genuine enigma, and that is all that I am concerned with indicating here.

'Numinosity'

There is an additional psychological phenomenon fundamental to the very notion of 'religion'. This is what the German Protestant theologian Rudolf Otto called 'numinosity', a term avidly adopted by C.G. Jung.[9] It refers to the sensation or experience of something being 'sacred' or 'holy' – a mingled feeling of awe, incomprehension, and the presence of power. It can be triggered by many things: places, persons, religious symbols, buildings, and music, for example. Numinosity is not to be confused with simple aesthetic responses; and although fear or dread may be one component of the experience, so may their opposites. As far as I know, although anthropologists may correct me, numinosity or something very like it is almost universally recognized in human cultures at all stages of 'development'. That which is numinous seems to possess its own energizing force and to be 'channelling' something *other* than unconscious contents of our own individual 'psychology', even though it is via this that it operates. The responses that it elicits are complex: most notably awed self-abasement, placation, or praise. Now, while it would be fairly easy to explain numinosity reductively in terms drawn from psychoanalysis (as Sigmund Freud did more or less explicitly in his 1919 essay 'On the Uncanny') or from Evolutionary Psychology, these do not, I feel, quite meet the case. We are dealing with what appears to be a basic human experience, which – even if relatively uncommon in its occurrence – lies at the heart of human culture and humanity's initial efforts to understand the universe and its place within it. It signifies not simple ignorance, but a general boundary or limit to our very capacity for understanding – yet in doing this it seems to constitute a form or source of knowledge in its own right. Numinosity should not be confused with 'mystical experience' or a feeling of 'oneness' with the world; on the contrary, its *otherness* is one of its defining features. We initially approach the numinous with fear and trembling. It is 'taboo'.

My concern here is not to interpret or explain numinosity, only to propose that it is a further, genuine, enigma which, whether we like it or not, and whether we have personally experienced it or not, we have to take seriously. It is probably the most immediate psychological source of all religion and concepts of superhuman divinity. (Although there are other kinds of source, numinosity strikes me as a necessary condition for these to begin operating, although clearly it does not explain religion's cultural forms and social functions.) And when experienced as enshrined in an individual person possessed of a certain kind of charisma, it becomes one of the most powerful factors in human history.

However, as I conceive it, numinosity can be malignant as well as benign or neutral. The most darkly numinous phenomenon of the twentieth century was surely Nazism, and the persisting common perception of the Swastika symbol as somehow intrinsically evil is, psychologically, quite remarkable. Its sheer power makes the Hammer and Sickle or Stars and Stripes look tame. Otto himself failed to acknowledge this negative form of numinosity. The topic is, it should be stressed, being very superficially treated here, my intention again being only to indicate the issue in question. Otto's intensely Christian account explores the role of numinosity in religious services, its modes of expression in art and music, and its evocation. I should not be understood as seeing it in quite the same way, as I am unconcerned with his pro-Christian goals.

Even leaving numinosity aside, both 'paranormal' or 'psychic' phenomena and the 'Mind in Nature' riddle have direct bearings on religion and religious belief, and on the boundaries of applicability of the physical sciences. This is reinforced by the way in which the former merge into the much wider category of 'religious experience'. We must, I think (slipping into the role of professional psychologist), take these issues seriously, if only because they are such powerful psychological *realities* for so many people. Given this, sceptical and dismissive atheist railing against religion is both futile and inadequate as a productive response to the issue. To assume that somehow physical science can or will account for all of them, dissolving all the puzzles to which they give rise, and that all the experiences in question are ignorant misunderstandings, is patronizing, dogmatic, scientific arrogance of the highest order, and not really a scientific position at all. That religion has as much evil as good in its history, and remains as ethically ambiguous as ever, is indisputable, but that is not our point here. Equal charges may be laid against science, and at least religion did not produce nuclear weapons, while secular ideologies have just as poor a moral record. Since religion is all-encompassing in its aspirations, attempting to frame all aspects of existence, perhaps its encompassing of evil is inevitable, whatever its protests to the contrary.

A brief note on psychology and religion

While, as has been repeatedly stressed, religion and physical science have become quite different in nature and purpose, any religion or philosophy of life must at least seek to reconcile its particular doctrines with the current state of knowledge of the physical universe. But, to use traditional language, religion is centrally about the 'soul' or, to use more modern terms, the 'psyche' or the 'inner self' – the 'mind', in short – in all its plenitude. This is where science – concerned with public, visible, physical phenomena – and religion inevitably encounter one another, and it is the boundary point where psychology emerged. It has since ranged from C.G. Jung's 'Analytical Psychology', in which religious issues are central and respectfully treated, to the explicitly reductionist and anti-religious work of the early behaviourists and, more recently, some neuropsychologists. In much contemporary psychotherapy and counselling, religious ideas of 'sin' and 'spiritual growth', for example, have, as previously noted, made a return to the discourse of psychology. Contrary to widespread popular belief, psychologists and religious professionals have, historically, actually collaborated or peacefully co-existed to a considerable degree, particularly in the field of counselling and psychotherapy.[10] To give just a single example, in Britain professional counselling was virtually created by the efforts of the devout H. Crichton Miller and Methodist minister Leslie D. Weatherhead in the 1920s and 1930s, coming to fruition with the establishment of the Westminster Pastoral Foundation in 1971, which was instrumental in the creation of today's British Association of Counselling and Psychotherapy and founded by another Methodist minister, William Kyle. This succeeded Weatherhead's City Temple Psychotherapy Clinic, founded in 1936. Psychologists have too often been described as modern 'scientific' replacements for ministers and priests. This does not bear much historical examination, but we must leave the point aside in this context, other than to observe that over the last century and a half religion (primarily Christianity) and psychology have interacted at numerous levels and in quite complex ways. We will return to psychology in the next chapter.

Summary

This chapter's basic argument is that there are indeed some genuinely enigmatic psychological phenomena, while the nature of 'mind' itself remains an unsolved riddle. It is unclear how orthodox natural science can tackle these issues, but it *is* clear that they are matters of central importance for our views of the world, our attitudes towards it, our philosophies of life, and the role of religion. As philosophical questions, neither the MBP nor the mind-in-nature issues obviously bear directly on the Creationism/ID controversy as such, but they do indicate that a hard party-line reductionist materialist strategy of attack is likely to be unsuccessful. Such a policy only lodges the controversy in the context of wider assaults on religion as a whole, alienating all but atheist materialists. This cannot be wise, and the enigmas that I have tried to indicate above actually put both of our parties on the spot. If reductionist natural science cannot tackle them satisfactorily, neither can literalist, Creationist, Christianity: it simply has no interest in such 'philosophical' issues. Yet covertly they provide the best route for finding a genuinely rational answer to the challenges of materialist scientific reductionism which so infuriate it.

As things currently stand, neither the puzzling phenomena and experiences referred to as 'paranormal', 'psychic', or 'supernatural', nor the persistent problem of the nature of mind, nor 'numinosity', constitute direct evidence for or against either Creationism/ID or its scientific opponents. At this stage what is important is not advocating any specific solution to such enigmas, but getting all parties to acknowledge their existence and importance. Finding a way forward will, somewhere along the line, entail paying them more concerted attention, but presently it is difficult to say how this should best be done, or who could best do it. It cannot, by its very nature, be done by using physical science's current methodological paradigms, but neither do literalist interpretations of the Bible (or any other major scripture) have anything to offer.

In conclusion, then, there *are* valid reasons for taking seriously the matters that we presently call 'religious' and 'spiritual'. And these stand, whichever side of the Creationism/ID controversy one is on.

8 Science as a religion

My main targets so far have been Creationism and Intelligent Design. Yet it would be most unfair to let science entirely off the hook. I have repeatedly asserted that physical science is not about providing a philosophy of life, and that it has a quite different role from that of religion. Unfortunately, as I willingly concede, scientists themselves have not always seen it that way.[1] Far from it. During the nineteenth century and into the first half of the twentieth, there were many who evangelically promoted science as a potentially superior replacement for religion. It was claimed that science was capable of providing a 'philosophy of life'. And this was nowhere more evident than in the case of evolutionary theory, where, indeed, such a view still persists in some quarters. The fervour of the strong Sociobiology camp which emerged in the 1980s has perhaps flagged somewhat, but its influence persists in sections of Evolutionary Psychology, for example, as well as in numerous popular science blockbusters. Dorothy Nelkin, a New York University professor of sociology, has drawn attention to the fervently religious tone of much sociobiological writing; she quotes its founder, E.O. Wilson, as actually saying 'Perhaps science is a continuation on new and better tested ground to attain the same end [i.e. as religion]. If so, then in that sense science is religion liberated and writ large.' She goes on to identify a plethora of recent popular science works, many evolutionary, with titles related to religion and 'philosophy of life', such as *The God Particle* and *Evolution and the Meaning of Life*.[2]

Its religious opponents were not mistaken in viewing evolutionary theory as a threat, given that many of its exponents enthusiastically used it to assail them – although it is rather pointless to ask who started the argument. While some, like T.H. Huxley, more or less confined themselves to attacking specific religious beliefs on scientific grounds, others, such as Herbert Spencer, Francis Galton, and many lesser lights, thought they could use the concept of evolution as a basis for a new philosophy of life which would indeed serve all of what they understood as the functions of religion. In a way they were returning to Romantic ideas common in Germany earlier in the century, when 'Transcendental Idealism' enjoyed enormous philosophical popularity. Whereas these previous thinkers, such as F.W.J. Schelling (1775–1854), had cast humanity's rise as a process by which some transcendental 'Mind' or 'Absolute' was realizing itself over time, the register adopted by Galton and Spencer was less sublime and more practical in tone. Although there was nothing in Darwin's own writings explicitly supporting the idea, evolution was readily presented as a progressive process, with humanity (especially white European humanity) at its peak.

This is actually a very tangled historical topic. Clearly in one sense nobody at this time, even Darwin, could have thought of evolution as non-progressive in some general fashion. What is true is that, whereas Darwin himself was extremely reluctant ever to enter into contemporary political debate, there were others like Spencer and Galton who leaped at the chance and used it to justify a self-aggrandising vision of Victorian Britain as representing the very pinnacle of evolution. Darwin himself usually kept his own counsel. For cultural reasons, the notion that evolution was somehow aimed all along at producing white humanity, which Darwin certainly would have rejected, rapidly infiltrated and permeated contemporary interpretations of what the theory was all about. The eugenics movement, the 'degenerationism' panic, and 'scientific racism', mentioned in Chapter 4, were the disastrous consequences of this. The idea that evolutionary theory can provide a basis for ethics and social policy while endowing humanity with a god-like significance continued to prove attractive to many scientists, notably during the 1980s, some of them sociobiologists. Humanity is on an ever-upward progressive curve, with science providing the means of ascent. It might be observed in passing that the very notion of 'progress' is now problematic in discussions of evolution, as well as in more general contexts.

As noted above, the titles of many recent popular science books, especially some of those published in the USA, suggest that they are contributions to religious and ethical themes. While this is obviously a good marketing ploy, it also reflects their authors' genuine assumptions that their grand 'scientific' visions do indeed illuminate such matters. Everyone is perfectly entitled to expound their visions of the meaning of life as they believe it has been revealed to them by their personal experience and knowledge – scientists being no exception. But being written by a scientist does not mean that a book is itself science. And when scientists market their science by giving the impression that they are contributing to religious thought, they cannot really complain when the religious promote religious texts as contributions to science.

There is one caveat to be entered here, however. Current understanding of the ecological complexity and inter-connectedness of the Earth's biosphere, combined with that of the Earth's geological and climatic history, led James Lovelock (1979) to propose that we should view it as if it were a living, self-regulating, organism, for which he proposed the name Gaia, after the Greek 'mother goddess'.[3] He has continued to develop this notion through a number of subsequent works. This suggestion was simultaneously a kind of scientific hypothesis and an ethical move. Lovelock's aim was to promote respect and responsibility for the natural environment by integrating the scientific and ethical bases for the 'Green' agenda. In doing so, he deliberately exploited the 'religious' echoes of the term Gaia, hoping to summon up something of the awe and reverence associated with religious belief. Nevertheless he explicitly stated that he was making an analogy with the notion of the goddess Gaia, not actually saying that the Earth *was* a person-like divinity, and he has been careful not to move beyond orthodox scientific understanding in the various relevant disciplines. Lovelock's unique position is perhaps as near as we have got so far to bringing religion and science into a genuine and satisfactory alignment.

Otherwise, while scientific knowledge may well provide a powerful source of inspiration for the religious, and in the past religion has certainly been a major motivating factor for many scientists, this does not imply that they are alternative versions of the same enterprise. From what has been said in previous chapters, this appears quite mistaken. Mary Midgley (a great Gaia fan, incidentally) has tackled the topic in depth in two books, *Evolution as a Religion* (1985) and *Science as Salvation* (1992), to which I refer anyone who wishes to explore the point further. Philosophies of life need not be religious, of course: humanism, for example, is explicitly non-religious, while some political ideologies, such as communism, have also served a wide range of needs traditionally seen as properly met by religion. The mistake is to imagine that science, *per se*, can serve these in any comprehensive fashion. What could a 'scientific' child-naming, wedding, or funeral be like? What comfort can science offer the grief-stricken? The only full-blooded attempt to create something resembling such a religion was Auguste Comte's Religion of Humanity in the 1840s and 1850s, which proved an utterly embarrassing failure. Religion is not, at heart, *about* acquiring knowledge of the physical world, even though all religions require a concept *of* that world.

If science now finds itself engaged in a battle with resurgent doctrines of Creationism and ID, it has, in some small part at least, itself to blame for how it is managing their encounter. I am not trying to exonerate the religious camp, but one core stumbling block is those in each camp who are unable to move beyond the terms of their mid-nineteenth-century perceptions and misunderstandings – both of each other and of themselves. The fairly strong line that I have been adopting here regarding the fundamental difference in character between religion and science does nevertheless need some qualification. As we saw in Chapter 1, the historical rise of Western science from the seventeenth century until the early twentieth century owed not a little to the support that it received from certain religious constituencies, and to the religious beliefs of many of its pioneers, not least Isaac Newton himself. It could indeed be argued that a belief in some version of Creationism was a necessary condition for science to get off the ground, for it both motivated such early pioneers and provided a rationale for assuming that the natural world was an orderly system amenable to rational investigation, rather than random chaos. Later British scientists similarly inspired included the Unitarian Joseph Priestley (1733–1804) and Michael Faraday (1761–1810), a devout member of the fundamentalist Sandemanian sect. Moreover, although the details of the relationship have been a matter of prolonged debate, most historians of science view the Protestant prioritizing of reason and personal experience over received authority as a central factor in promoting widespread receptivity to scientific ideas, especially in the seventeenth and eighteenth centuries. Nevertheless, as we have been insisting, while a religious belief in divine Creation or Design might have provided the *motivation* for engaging in proto-scientific work, these beliefs did not constitute explicit 'scientific' theories or hypotheses guiding such scientists' actual research, although of course religious beliefs of any kind might favour thinking in a way which proves scientifically useful – as with Faraday (for more on this, see G. Cantor, 1991).

Much is made of this close relationship by the sociologist and historian of science Steve Fuller in his controversial *Dissent over Darwin* (2008), in which he argues for the continuity of design orientations in mainstream science and the legitimacy (if not correctness) of the contemporary ID camp's challenges to Darwin. His error, as I see it, is a failure to acknowledge that the eighteenth- and early nineteenth-century version of the Argument from Design was radically different, both in purpose and in logic, from the way it is used today (as argued in Chapter 2 of the present work). Fuller rather obfuscates this by taking any casual use of the term 'design' as signifying acceptance of the Argument from Design perspective: thus, for example, to say that penguins' feet 'are designed to withstand extreme cold' is enough to place one in the ID camp. This will not do. The central point at issue was how to explain the 'adaptation of means to ends' in 'animate Nature': eyes, wings, leaves, lungs, roots all appeared to be perfectly designed to serve their biological functions. Despite its shortcomings, as detailed in Chapter 2, the Argument from Design appeared to many as really the only option on the table, and certainly the most awe-inspiring and satisfying one, especially given that the Creation was accepted as happening only a few thousand years ago. It was the general abandonment of this assumption that caused the downfall of the classic Argument from Design, as other more impressive options now joined it in the light of continually extended time-scales and proto-evolutionary biological theories such as Lamarck's and Erasmus Darwin's. Hostility to religion, if widespread in nineteenth-century science, was not the primary cause of this change in opinion. The Argument from Design did not succumb to some anti-religious, materialist, atheist attack: it simply withered as it lost credibility and scientific utility. The famous notion of 'warfare between science and religion'[4] bears little close examination, except in some fairly restricted contexts. To pursue this any further here would involve too great a digression.[5]

In the twentieth century there continued to be a tradition, or genre, of works written by eminent scientists in which they strove to recapture earlier affinities between the scientific and religious visions of the universe as wondrous, awe-inspiring, and mysterious. These works have included James Jeans's *The Mysterious Universe* (1930), Alister Hardy's *The Living Stream* (1965) and *The Divine Flame* (1966), and Fritjof Capra's *The Tao of Physics* (1972). Also in this category we could place many of the works by the Jesuit theologian and palæontologist Teilhard de Chardin, the most popularly successful being *The Phenomenon of Man* (1959).[6] And, as already mentioned, the flow of popular science texts with quasi-religious aspirations persists unabated. I can only repeat that whatever their sometimes considerable merits, such works are not themselves *science*, but rather draw on science as an inspiration or resource for addressing essentially non-scientific questions which are the proper province of religion, philosophy, and literature. Visual artists and composers have similarly found inspiration in science, among innumerable other sources.

So, while modern science was historically rooted in religious (more specifically Christian) concerns, and while scientists have sometimes viewed it as potentially replacing religion, the two are at heart addressing quite different kinds of puzzle. (Islamic concerns also of course informed the earliest phases of European scientific thought and were crucial in many respects – not least in the invention of algebra and replacement of Roman numerals.) As always, there are fuzzy cases, but these do not invalidate the basic picture that I am proposing. Teilhard de Chardin, for example, strove to formulate an integration of theology and evolution, in which he saw evolution heading towards a divine 'Omega Point'. Lovelock's 'Gaia hypothesis', referred to earlier, is another special case.

One argument which remains unsettled (even in the sense of whether it is a sensible argument in the first place) is that we need to explain why the universe is comprehensible to humans – the 'anthropic' principle espoused by several eminent scientists. I am unclear whether the direction of the argument is that the universe was designed to be comprehensible to humans, or that humans were designed to be able to comprehend the universe. Since even to identify something as 'incomprehensible' is to have taken the first step towards understanding it – i.e. by naming it and circumscribing it – this strikes me as all rather tautological, since that which is truly incomprehensible is something of which we would not even be conscious, and there is nothing to say that phenomena in this category do not exist in the universe. Enough.

The most problematic discipline is actually psychology, much of which is concerned with matters of mental distress and, more broadly, the nature of human fulfilment. Many psychological topics are impossible to discuss in a value-free fashion, for the topics themselves, and the language for referring to them, arise within ethical contexts. We cannot discuss child development without some evaluative notion of what the psychologically healthy child should be like, and 'moral development' – how a child's ideas of right and wrong are formed – is a further intrinsic part of this. In personality theory, all kinds of terms like 'authoritarian' and 'open-minded' are implicitly value-laden. But although many psychologists have been devoutly religious, very few, including the numerous ordained Roman Catholic priests who figure in its history, have (at least overtly) sought to use the discipline to legitimate their specific religious beliefs, albeit their approaches are implicitly framed within the terms of underlying theological doctrines (in the case of Catholics, usually Thomist).[7] Even C.G. Jung, who identified the psychological plight of 'modern man' as due to a loss of religious faith or grounding, was more concerned with 'scientifically' (as he considered it) analysing and articulating this, than with endorsing a specific religion.

The rise of science may be viewed as a kind of division of labour which emerged from the all-encompassing Christianity which preceded it and which, like all such traditional religions, *did* offer explanations of both physical and spiritual existence, without actually clearly differentiating them. All could be included in a single, unchanging, religious cosmology. In time, however, it became clear that scientific understanding of the physical world was fundamentally different, both in kind and in use, from theological 'knowledge'.

This raises an important issue which, if something of a digression, must be addressed. The late nineteenth-century ceasefire between science and mainstream Christianity was largely achieved by Christians accepting the literal truth of scientific knowledge but insisting that their own knowledge was equally valid as 'spiritual' or 'symbolic' truth. The difficulty is that in a genuinely unified religious cosmology there has to be a belief that, however rich, sublime, symbolic, and spiritual its meanings can be, they are as it were 'ratified' by a base of literal truth. One could even argue that technically the concept of 'symbolic truth' is nonsense, because truth is something which applies to propositions, and symbols are not propositions. This, though, is too narrow a statement of the problem, since, as pointed out previously, the very richness and enduring value of the world's major religious texts (and great literature, art, music, and drama also) do rest on their being experienced as 'true', albeit 'symbolic', reflections and expressions of the human condition. But if there is *no* level of literal ratification for religious doctrines, they can only possess the same status as these other enduring genres – great manifestations of the human spirit possessing profound psychological 'truth', to which as individuals we may or may not respond, but with no additional authority such as religions have to claim. The fundamentalists have intuitively understood this in a way that many Christian intellectuals have not. Symbolic and spiritual truths alone are not enough: there has to be a concrete empirical level as well.

The question then arises as to which biblical content *does* have to be taken as literally true, as well as 'symbolically' true, for the believing Christian. As I am not a Christian, it is hardly my task to answer this, but it does appear to me that it really boils down to the events from the run-up to the Crucifixion to Paul's conversion on the road to Damascus. If the biblical account of these events, and their meaning for those who witnessed them, is not fundamentally true, then the whole edifice crumbles. Their subsequent theological interpretation (there is little explicit theology as such in the Bible) or Christ's actual status as Man, God, God-in-Man, literal Son of God, and so forth are matters on which Christians have never achieved universal consensus: matters, moreover, which have caused much internecine bloodshed among them over the centuries (not least those immediately following the founding of Christianity). The literal truth of the Virgin Birth and Nativity story is expendable. The miracles may well be accounts of real events, but they are not unique in comparative-religion terms, similar achievements being ascribed to numerous religious and spiritual figures, particularly in the Indian subcontinent. As far as the Old Testament is concerned, for Christians (though not Jews) its main value is surely in setting the religious context in which Jesus worked. While incomprehensible without the Old Testament, the truth of Christianity is hardly dependent on the Old Testament's literal truth. It is, to repeat, actually so variegated in its content – from poetry (*Song of Solomon, Psalms of David*) to political and military history, prophecy, accounts of tribal or ethnic origins and genealogy, philosophical parable or meditation (*Ecclesiastes, Job*) and, in my view, typical ancient mythologizing (especially *Genesis I–IV*) – that academic lifetimes have been spent attempting to unravel it. Which, of course, is why it is so uniquely remarkable.

Things are made additionally complex in this case because it is clear that the texts comprising it, especially the earlier ones, underwent numerous politically and religiously motivated censorings, re-editings, and rewritings before acquiring their present form (see also Appendix B). (One might add that the inclusion of *Revelation* in the New Testament, in 397 C.E., has always been considered somewhat questionable, and hence Luther found it rather problematical and the Orthodox churches do not include it in scripture readings during services.) While much of the Old Testament's purely historical content is probably basically true, and often (if not always) archæologically supported, Christianity as such surely does not depend on this. It almost certainly, however, gives a misleading impression of how far back strict monotheism dated. Traditionally Christianity has, as I well understand, cast the Old Testament in its entirety as a divinely inspired documentation of a plot-line culminating in the Incarnation of Christ. Even so, this leaves open the extent to which it should be read 'symbolically' or 'literally'. My points here are really twofold: firstly that a religion has to make at least some core claims to literal truth if it is to be believed to be what it says it is; and secondly that in the case of Christianity the only really necessary literal-truth claims appear to me to relate to Christ's last days and their immediate aftermath. On the other hand, were even these literal claims to be empirically, 'scientifically', confirmed, belief in their truth would, paradoxically, remove them from the realm of religious *faith*, as we saw earlier – which is not, presumably, what Christians want.

The notion of theological 'knowledge' is also, admittedly, rather problematic, since there are, I think, no concrete knowledge claims on which theologians agree; but on the other hand, knowledge *of* theology can, I think, be valuable, and some theology at least is of enduring merit as a source of wise meditation on the human condition, even if the nature of its official subject matter, God, remains as obscure as ever despite all its efforts. Digression over.

Conclusion

The relationship between science and religion is a major topic in its own right, on which vast amounts have been written, but I hope I have said enough here to indicate that (convoluted though the story may be) they have become quite different kinds of human enterprise, offering quite different kinds of knowledge.

To reinforce the point, a final, very significant, difference between them should also be stressed. Science still remains basically unified: whatever their theoretical disputes, and regardless of the variety of the things that they study, all physical scientists share a common understanding of the over-arching enterprise in which they are engaged. Religion is, by contrast, inherently diverse. There is not one 'religion': there are numerous religions and numerous divisions within each of them. Their beliefs and creeds are frequently incompatible with each other, and when push comes to shove their conflicts have regularly erupted into bloodshed: Protestantism versus Catholicism, Shi'a versus Sunni, Hinduism versus Islam, and Christianity versus nearly every other faith it has bumped into. Subordinating themselves to religious, ideological, and nationalist causes, as they so often do, conflicts between scientists are generally not about science as such. The exceptions have been when scientists have felt that fellow scientists have allowed their *scientific* judgment to be over-ridden by ideology, as in the cases of Nazi race theory and Soviet-era Lysenkoist genetics under Stalin. The scientific merits of Soviet physics and Nazi aeronautics, by contrast, were not in dispute. It is also possible to be a scientist *and* have profound religious convictions, yet you cannot have two equally profound religious affiliations, being, for example, a Christian *and* a believer in Judaism, a Muslim *and* a Hindu. *Not* being a religion is now actually, albeit covertly, one of science's defining characteristics.

But the real twist is perhaps that, by virtue of their fundamental failures to grasp the core differences between religion and science, what American Creationism and ID are actually trying to do is convert religion into science, rather than convert science to religion, at the same time as some in the scientific camp are still trying to do the reverse! Such aspirations would therefore appear to be putting each of them on a course for inadvertent self-destruction, were they to be met.

9 Moving on

If we are to move beyond the sterile confrontation between Creationism and ID on the one hand and fervently materialist physical science on the other, we must, I suggest, adopt the more psychological tack discussed earlier. This is, after all, a confrontation between certain groups within the broader religious and scientific worlds, and many in each group wish to disown the entire contest. Moving on will not involve the unambiguous and permanent triumph of one side over the other, for the covert psychological roots of their conflict are probably perennial. But if we can render these roots more accessible to intelligent scrutiny and analysis, we might be able, at least, to shift the debate to a more sophisticated, collaborative, and fruitful level.

In this short work all I can do is to make some provisional suggestions. And to start with we could do worse than ponder on Paul Tillich's 'three types of anxiety', around which his work *The Courage to Be* (1952) largely revolves. Although often couched in an existentialist terminology which some may find a little inaccessible, Tillich's message is at heart fairly straightforward. There are, he proposes, three kinds of anxiety, corresponding to three ways in which we feel that our existence can be threatened. The first is 'anxiety of death', the second is 'anxiety of meaninglessness', and the third is 'anxiety of condemnation'. These three forms of anxiety are natural and inherent aspects of human existence: there is nothing pathological about them. It is when we are unable to manage them, when our courage yields to despair, when one or other of these anxieties threatens to overwhelm us with 'nothingness', that they can become 'neurotic' or dysfunctional. (They are of course inter-related, and Tillich further identifies various modes in which each can manifest itself.) Of special interest in relation to Creationism/ID are some of Tillich's observations on the second of these: 'anxiety of meaninglessness'. While a certain kind of doubt 'is a condition of all spiritual life', he observes that '(t)he threat to spiritual life is not doubt as an element but the total doubt'. If this becomes sufficiently intense and overwhelming, then a person tries to break out of this situation, to identify himself with something trans-individual, to surrender his separation and self-relatedness. *He flees from his freedom of asking and answering for himself to a situation in which no further questions can be asked and the answers to previous questions are imposed on him authoritatively.* ... Meaning is saved but the self is sacrificed. ... Fanaticism is the correlate to spiritual self-surrender: it shows the anxiety which it was supposed to conquer, by attacking with disproportionate violence those who disagree and who demonstrate by their disagreement elements in the spiritual life of the fanatic which he must suppress in himself. His anxiety forces him to persecute dissenters. (pp. 55–7, my italics).

Tillich goes on to note that this can happen when, as in twentieth-century Christianity, the power of traditional doctrinal symbolism weakens and current conditions become remote from those in which spiritual doctrines originated, ceasing to be able to satisfy present needs. This relates back to the numinosity phenomenon discussed in Chapter 5, the weakening of the power of traditional symbolism, meaning that it has lost its numinosity, its immediately felt aura of 'holiness'.

One does not, I think, have to be a Christian or an existentialist to agree that this analysis casts considerable light on the psychological roots of religious fundamentalism and the fanaticism often associated with Creationism.

If Creationism is rooted in the fear of meaninglessness, from what, in Tillich's scheme of things, might scientific 'fundamentalism' of the kind espoused by Richard Dawkins and many other contemporary anti-religious, often atheist, scientists derive? Perhaps from the third type of anxiety: fear of condemnation. The physical sciences enjoyed a prolonged triumphant history from the early seventeenth century down to the middle of the twentieth. As 'reason' incarnate they, and their technological consequences, have transformed human existence, first in Western cultures, and now globally. From transport and manufacturing methods to medicine and communications, from architecture and the storage of information to the materials in which we clothe and feed ourselves – no aspect of human life now remains untouched, least of all how we fight wars. And most of the changes thus wrought appeared initially to be of unalloyed benefit, raising the quality and extending the length of life and the range of experiences to which ordinary people had access. Military technology was always perhaps the proverbial fly in this ointment; but as long as ensuring national military prowess was considered a virtue in itself, this could be overlooked. Whatever the sins of other professions, scientists could enjoy a clear conscience. By around 1900 science was widely viewed as the guaranteed route towards solving all humanity's problems, with the publication of numerous utopian visions of what it might yield, a view shared by the majority of scientists themselves. As described in the previous chapter, science itself could become a 'philosophy of life', replacing religion. Scientific salvation was at hand. And – ironically perhaps – the Great War of 1914–1918 initially only reinforced this, the catastrophic disaster in effect amounting to a burning of all bridges to the past, leaving science alone offering a credible route ahead (in some cases taking the form of an ideological commitment to Marxism as itself an apparently 'scientific' political and economic philosophy).

The last seventy years or so have, however, seen the moral unassailability of science called increasingly into question, even as its achievements continue to escalate. Following the first testing of the atomic bomb, Robert Oppenheimer is famously reported as saying 'the physicists have now known sin'. But it is not only advances in military uses of science which have damaged its image: it now seems that on virtually all fronts the intended and unintended consequences of science-based technology are proving horrendously problematic. These require no detailed rehearsal here, but they span the spectrum from global warming, damage to the ozone layer, and carbon emissions, to the problems of disposing of plastic and nuclear waste, industrial pollution, and the vanishing of the Aral Sea. Science's efforts to stay ahead of the game by constantly holding out the promise of a further impending technological fix are met with increasing scepticism, and even within its ranks entire sub-disciplines like climatology, oceanography, and ecology are shifting into the ranks of the doomsayers.

For those scientists who have centred their lives and identities on what they have romantically viewed as the morally impeccable, disinterested pursuit of truth, the present situation must surely then cause some 'anxiety of condemnation'. That the scientific cosmology, evolution included, is intellectually, aesthetically, and in some sense perhaps even spiritually, wonderful is not in question. What is problematic is the mayhem that ensues in the wake of scientific knowledge. Every scientific portrayal of the wonders of the terrestrial natural world now comes with a downbeat health warning about how the wonder in question is under threat. Science, the very project that revealed such wonders, has, it seems, endangered them in the act of doing so. Philosophers of science have long argued that it is, logically, impossible to observe phenomena without affecting them. In practice, most scientists initially felt that, outside nuclear physics, they could ignore this argument for practical purposes (although psychologists did begin to worry about 'experimenter effects').

Now, however, the implications of this insight are being writ horrendously large. Dismissing the negative consequences of science as 'misuses' is even less convincing than producing excuses for the sundry crimes committed in the name of religion, not least because science has been substantially funded by the very agencies that have brought them about. In psychological terms, genuinely idealistic scientists, whose life's meaning is centred on their faith in their calling, may be unable to face up to the ever-more evident moral ambiguity of the entire project. Here are some observations made by Tillich about the fear of moral condemnation.

> Within the 'contingencies of his finitude'... (a man) ... is asked
> to make of himself what he is supposed to become, to fulfil his
> destiny. In every act of moral self-affirmation man contributes to the
> fulfilment of his destiny, to the actualization of what he potentially
> is. ... But ... man has the power of acting against it, of contradicting
> his essential being, of losing his destiny. ... *A profound ambiguity*
> *between good and evil permeates everything he does, because it permeates*
> *his personal being as such.* ... The awareness of this ambiguity is the
> feeling of guilt. (p.59, my italics).

One response to this is 'moral rigour and the self-satisfaction from it', or 'legalism' (p.60).

While less extensive and somewhat more tangential to the issue than his observations on the fear of meaninglessness, this passage offers some food for thought. It certainly helps to illuminate the way in which scientists often wriggle legalistically when confronted with accusations of moral irresponsibility, and the morally self-righteous tone of some who claim to speak for Science with a big S.

As far as the physical universe is concerned, I must stress that I find the current scientific cosmology, including evolution, infinitely superior both aesthetically and intellectually to that offered in *Genesis*. One may in fact search the Bible in vain for any example of concrete knowledge about the nature of the physical universe. This is not surprising, since providing such knowledge was never the intention of any of its numerous authors. But whatever the merits of the scientific cosmology, it is now clear that contemporary science is in no position to occupy any moral high ground. Knowledge is, proverbially, power, yet science has been studiously, even stupendously, indifferent regarding to whom and for what purposes its knowledge has been dispensed. Until the mid-twentieth century it regularly washed its hands of moral responsibility, on the grounds that the applications and uses of its knowledge were matters for 'society' at large to decide – even as it lamented general scientific ignorance. Like institutional religion, in fact, science largely earned its living by supporting the four great causes: war, health, food, and wealth. The only difference between religion and science in these respects was that calls to the Almighty to smite national enemies proved less effective than calls to the scientists to devise weaponry to do so, and that, whatever their psychological benefits, prayers for recovery from disease proved less effective than antibiotics.

In some ways the moral position of science is in any case less secure than that of religion, for while religion views morality as essentially an individual matter which contains within it procedures of moral redemption for whatever moral failings its believers and institutions are guilty of, either individually or collectively, the morality of science is narrowly located within an overarching ideology which casts it as the embodiment and vessel of pure and objective reason. When professional participation in the scientific project becomes a scientist's moral centre, challenges to the credibility of this ideology (especially, I must add, apparently 'scientific' challenges) must prove highly disturbing to those for whom it plays this psychological role.

What I am suggesting, then, is that both religion (most obviously, but not uniquely, Christianity) *and* the physical sciences find themselves at a point of profound psychological crisis. A full analysis of the nature of these crises is urgently needed, although I cannot offer one here. I do, however, suggest that a useful starting point could be an exploration of how far Tillich's concepts of the fear of meaninglessness and fear of condemnation might bear on them. But while this may help at an intellectual level, it is hardly likely to have any quick impact on the ground. This requires more direct attention to immediate issues.

Types of Creationist and ID supporter

One question which needs to be confronted is whether advocates and supporters of Creationism/ID genuinely believe in what they are saying. There is no general answer to this, but we can perhaps identify three broad groups. First there are honestly devout Christian intellectuals, often with scientific training but affected by the kinds of psychological conflict that we have just been considering. These do indeed believe that there are basic loop-holes in the received scientific cosmology which can be rectified only by a reconciliation with the ID position. Secondly there are, in the USA at least, a very large number of lay Christians for whom the Bible's authority necessarily over-rides that of science in all matters and who are thus also, as they interpret it, honestly bound to accept the Creationist account as found in *Genesis*. Were these the only constituencies in play, there would be relatively little to worry about. The first can be engaged, in principle, by theological, scientific, and philosophical debate, while the second would be likely to shrink, as it did in Europe, in the face of broader scientific education. There is, however, a third group. These are those whose real agendas are political and ideological, typically on the right of the US political spectrum, who see the controversy as exploitable to promote these other ends, which have nothing to do with scientific knowledge. Rather oddly, they cast 'Darwinism' as a manifestation of the 'liberalism' that they detest – a peculiar reversal of the ideological connotations that the theory of evolution acquired among liberals in the mid-twentieth century as having been misused to provide a rationale for fascism, eugenics, and genocide. There is, to be candid, something quite opportunistic about American right-wing anti-evolutionism. Historically, ever since the early 1900s, it has been from liberals and the left that attacks on the use of evolutionary ideas as a framework for understanding contemporary human affairs have usually emanated. And ever since Herbert Spencer, it has been the political right that has been keenest to utilize them as a justification for competitive capitalism, anti-internationalism, racism, and opposition to feminism. Supporting or expressing sympathy for Creationism/ID enables the political right to gather electoral support from the fundamentalists and pro-ID academics. But you cannot convincingly blame 'Darwinism' for racism *and* see it as part of a plot by the historically anti-racist liberals!

This 'doublethink', to use George Orwell's phrase, engenders and expresses a dangerous ambivalence towards science in general, rooted in the fact that on balance scientists have tended towards politically liberal positions and are, being independently minded intellectuals, in any case suspect in the eyes of those who see themselves as tough-minded patriotic denizens of the 'real world'. This last point is demonstrated by the peculiar mix of support and rejection that science received under the presidencies of Ronald Reagan and the two Bushes.[1] When geology claims to be able to locate oil deposits, it is heard with enthusiasm; when it says that the Earth is millions of years old, it is kept at arm's length; meteorological skills in weather prediction are avidly integrated into national life at all levels, from the military to tourism and farming, but when meteorologists conclude that global warming is a reality, then the hunt is on for minority dissidents from this overall disciplinary consensus. And while physicists and astrophysicists can obtain endless funds for military-related research, if they also claim that the universe was created in the Big Bang, well, that is regarded as mere speculation. Such ambivalence is common to most regimes but is glaring in the US case. I hasten to add that there is nothing odd about this: it is only a particularly dramatic example of the universal tendency to believe information which supports our existing views and to disbelieve that which is at odds with them, regardless of the weight of evidence. But this is a fairly commonplace piece of wisdom, and it does not excuse those in power for making the same error.

To the extent that Creationist/ID doctrines pose a genuine threat, rather than simply being manifestations of that delightful human capacity for pig-headedness and eccentricity which contributes so much to life's rich variety, it occurs when they reflect this kind of intellectual irrationality.

Naturally it must be conceded that if you genuinely believe that you are fighting Satan, you are not inclined to be too fussy about the methods that you use – and I do not doubt that genuine Creationists and ID supporters are often happy to eschew the rules of fair play. This would matter little if the third camp did not provide them with such ample resources for doing so. It is the opportunistic co-option and promotion of these ideas by neo-conservative and other right-wing political ideologists which I find disturbing, along with the fact that in Britain this covertly enabled them to gain a foothold of respectability in primary and secondary education under the Blair government, which felt bound, for populist reasons totally unrelated to education, to lean over backwards to support 'religion'. As noted in the Introduction, this situation has subsequently changed somewhat, but the underlying sentiment persists.

One move which needs to be made, then, is for someone to undertake research akin to that done by William Tucker (1994) and others on the funding of psychological research aimed at proving race-related differences in intelligence, and other racist beliefs.[2] This is necessary in order to pin down, and subsequently publicise, precisely where the real money is coming from.

More broadly, however, moving on must eventually entail a basic confrontation over the very natures of those activities that we currently call 'religion' and 'science', and their roles in society: a confrontation not only between two entrenched camps but between the various parties within each of them; and it should also engage those who wholeheartedly identify with neither (and who in Britain may even amount to the majority). How near to this we are, I cannot guess.

Moving on in education

Faced with a world-view that is fundamentally incompatible with one's own, there are three possible solutions: physical annihilation of one's opponents – which is wicked; their conversion – which is, in this case at least, unachievable; and mutually tolerant co-existence – which, however difficult to manage, is the only viable strategy. As far as education, especially moral education, is concerned, one central aim must be to convey an understanding of the necessity for such tolerant co-existence, without hiding the problems. 'Tolerance' does not of course extend to tolerance of dishonesty, injustice, deceit, law-breaking, prejudice, and outright evil. One problem, however, is that certain world-views can appear to legitimate the committing of one of these sins 'for the greater good' – especially when combating what is believed to be one of the others. The difference between engaging in civil disobedience in opposition to unjust laws or policies and bombing abortion clinics in opposition to what is believed to be an outright evil may seem clear enough to most, but between these examples are innumerable grey areas. Beliefs have consequences for action. This does not, however, alter the fact that mutual tolerance should be the default position.

When it comes to specific curriculum subjects, society is, I think, bound to require a teaching of the 'truth' as currently understood by acknowledged experts in the field. Thus we do not give Holocaust denial a sympathetic airing in History, do not raise the possibility that the Sun orbits the Earth when explaining basic Astronomy, do not pretend that continental drift is still in doubt in Geography, or ascribe *Hamlet*'s authorship to Francis Bacon in English Literature. Similarly, there should be no obligation to teach Creationism and ID in science classes.

That is not to say that the existence of such views should remain unmentioned, but the approach should be in terms of rationally explaining why scientists consider them fallacious and have thus rejected them. This should not be done in an oppressive or emotionally hostile way. Pupils who fail to toe the line should certainly not be punished or otherwise negatively affected for so doing. And science teachers should not hold forth about the Bible itself, any more than teachers of religious studies should pronounce on scientific questions. More specifically, the treatment of Creationism and ID should explain that the issue is not one which has arisen *within* science itself due to some crisis in evolutionary theory or cosmology inside the various relevant disciplines themselves; nor even has it arisen within academic theology as a response to a crisis in *that* discipline: as argued in Chapter 6, its revival can mainly be attributed to the recent access to economic and political power of a fundamentalist constituency located in the Mid-West and Deep South of the USA, a constituency with a somewhat anachronistic sub-culture which long isolated it from, and rendered it hostile to, the rise of modern science elsewhere in both the United States and Europe. To be against Creationism and ID is not to be 'anti-American', but it is to resist certain irrationalist waves emanating from this particular component of its present culture.[3]

Such an analysis could be more effective in countering Creationism and ID than standard 'scientific' refutations, which perhaps yield too much by treating it seriously in orthodox scientific terms, and would at the very least provide a useful complement to these. It would be additionally helpful if non-Creationist teachers of religion could explain why Creationism and ID are a religious *cul-de-sac*, since by seeking to emulate physical science they represent a complete departure from religion's central concerns with the individual's spiritual and moral well-being (the very things that earn it a prominent social presence), while also, as we saw, covertly negating the core character of religious Faith.

In the present author's view, public education should be resolutely secular, which is not the same as being anti-religion. The current UK requirement for a 'daily act of worship' to be organized in schools, already more honoured in the breach than in the observance, should therefore be dropped. On the other hand, banning the celebration of the major religious festivals associated with pupils' various faiths, and obsessional policing of the wearing of clothes and jewellery with religious connotations, strike me as miserable and humourless gestures, pandering to the purely hypothetical possibility that they may 'offend' intolerant believers of one creed or another. That is the very opposite of what public education should be doing: combating intolerance. An education which omitted any coverage of religion would clearly be incomplete, but its coverage should be presented from a neutral standpoint, precisely because there is no expert consensus on the validity of any of its contents, from the existence of God (or gods) downwards. The continued, government-encouraged, growth of 'faith-based' and faith-sponsored schools does not augur well, however.

But it was not my intention to deliver a sermon. Any ideas that I have about 'moving on' are quite modest, but sorting out the sheep from the goats in both the mainstream physical science and Creationist/ID camps would be a good start. As would setting our educational house in order.

Finally, I hope I have helped readers to get the nature of the Creationism and Intelligent Design issue into focus. It is clearly a matter of considerable complexity. This does not mean that someone intelligently designed it to be that way.

CODA

In the foregoing I have put forward several arguments against Creationism/ ID which I believe have rarely been properly articulated or aired. The debate generally hinges on a restricted range of issues, usually pertaining simply to empirical evidence. While this central ground was discussed in Chapter 4, these others may usefully be summarized as points which Creationists have to answer.

- Complexity as such does not signify design, and in the case of human artefacts it usually signifies a plurality of designers and a high degree of division of labour in their actual manufacture. In most cases no single person actually has a comprehensive understanding of the workings and manufacture of such artefacts, which are, moreover, in a constant state of development or 'evolution' over time. So much for Paley's watch.

- The whole point of religious *faith* is precisely that it transcends reason (even though rational arguments are frequently deployed to support it). There is no virtue in believing something simply because you know it to be true. (Since science explicitly relies entirely on reason and empirical experience, it may perhaps be said to have a faith in Reason, but the use of the term here is somewhat metaphorical, and this does not make it a religion.) To try to prove, scientifically, that the Bible is literally true is thus to try to remove it from the realm of religious faith altogether. Religious conversion is a choice, and often a matter of prolonged and tortured inward struggle. The scientist has no choice about the facts – although he or she may have a choice about deciding what exactly they are. Religion is a way of life. Being a scientist is a career – although, like any career, when ardently embarked upon it entails a particular life-style. And while anyone can call themselves a Christian or a Muslim or claim any other religious appellation, to be a scientist requires a formal qualification. (There is a certain irony in the following quote that I found on the Catholic website referred to in Chapter 1 note 18: 'On the one hand, there are scientists who appeal to evolution as evidence to reject faith in God. This is wrong. But, on the other hand, the attempt to turn religious faith into science is the same mistake, but in the opposite direction. Both creationism and intelligent design fail in this regard.')

- Hence, for the believer, the 'evidence' which supports religious belief is fundamentally in the form of subjective psychological experiences (for instance, awareness of a comforting divine presence), but the evidence supporting scientific belief is in the form of publicly observable phenomena and the practical efficacy of theoretical formulations of their character.

- The puzzles that religion and physical science aim to solve are thus of completely different kinds. The former are, to put it crudely, about the meaning of human life. The latter are about how the physical universe works.

- The enduring value and survival of religious scriptures, as of all great literature, do not rest on their literal truth. Works which contain only 'facts' as currently understood are inherently transient, and hence, barring other qualities, scientific works rapidly become obsolete and largely unread. Nor are they intended to be anything else. Science is generally intended as a guide to practical engagement with the physical world; religion and literature are intended to stand on their own intrinsic merits as sources of insight into, or explorations of, the human condition.

- It was noted towards the end of Chapter 2 that there appeared to be a conflict within Creationism between those inspired by the Argument from Design and those insistent on the literal truth of the *Genesis* account. Although the existence of these variant positions has long been recognized, with the latter viewed as more 'dogmatic' or 'fundamentalist', on closer scrutiny they are surely fundamentally opposed. The former sees the Universe as a beautiful and awesome divinely engineered phenomenon to be praised; the latter sees it as transient and inferior, having, since the Fall, become the domain of Satan and due for imminent demolition. These two positions do not, in other words, differ simply in degree. We might legitimately ask for some clarification on this from the Creationism/ID camp, which has so far brilliantly succeeded in papering over this internal schism.

- Historically, the nature and origins of contemporary Creationism can be largely explained by fortuitous sociological circumstances which isolated extensive regions of the United States from direct contact with mainstream religious, scientific, and cultural developments. As a result, fundamentalist religion was able to endure as a kind of 'fossil survival' in forms which had largely disappeared elsewhere.

We may therefore demand to know, more broadly, why it is that Creationists/ ID supporters are so intent on striving to turn their religious beliefs into a science, when that would entail losing everything genuinely religious about them?

The physical-science camp is not, as I have indicated, beyond reproach. It has its own tendencies towards self-inflation into a religion. It has a dismal moral record, especially over the last hundred years, comparable in iniquity to anything in the history of religion. And it can exhibit arrogant and patronizing contempt for religion in general. It has also been woefully complacent and unimaginative in its responses to Creationism, largely by allowing its opponents to set the terms of the controversy to their own advantage. Those in this camp have thus left potentially fruitful alternative approaches, such as those of social history, sociology, psychology, and critical textual analysis, largely unexplored. Physical science has also, surely mistakenly, often assumed that reductionism can be extended into the understanding of human affairs, occasionally 'reducing' or dismissing even consciousness itself as an 'epiphenomenon' of no real consequence. The anti-religious vehemence of some scientists also appears to be a defence mechanism against their own personal moral anxieties. One can, however, be unambiguously anti-Creationist without being anti-religion in general.

I hope that the present work will go some way to widening the debate and forcing Creationists to address aspects of their position which they have hitherto been able to ignore. I also hope that it may prod some of their opponents into developing wider and wiser ways of engaging with them.

Appendix A: Understanding fossil evidence

I cannot claim any special expertise regarding fossils, but I am aware that a number of misunderstandings, presented below, appear to be in circulation among supporters of Creationism and Intelligent Design, or are perhaps simply exploited by some of the less scrupulous.[1] My only serious foray into this field was in connection with background research for my book *Human Evolution. An Introduction for the Behavioural Sciences* (1987). I doubt that Creationists will find that fact very reassuring.

'Fossil gaps'

Some Creationists still assert that the fossil record has major gaps, and that this is particularly so with regard to 'transitional' fossils, intermediate between species. This was a widely used argument in the nineteenth century, but it fell from favour as the gaps kept getting filled. Even so, we do perhaps need to look at the topic again, if only to avoid being accused of dodging the issue. That there should be gaps is in itself hardly surprising, since fossilization is an extremely rare event, except under certain marine conditions, at least in a way which leaves the form of the organism intact rather than turning it into oil, coal, or chalk (for example). There is actually a sub-discipline called 'Taphonomy' which studies what happens to organisms after they die. Interested readers are advised to check this out. Pat Shipman (1981, reprinted 1993) is as good a place to start as anywhere.[2]

The fact is that the environmental conditions in which fossilization can occur are very far from universal, and in the majority of circumstances virtually no fossils will be left at all. Factors such as soil acidity, scavenger species, prolonged surface exposure, and subsequent erosion of beds in which fossils were deposited are all involved. This means that the fossil record is hardly representative, since there will be gaps in terms of species which lived in environments unfavourable for fossilization, or where subsequent erosion destroyed them. If an intermediate species lived in one of these, it would leave no fossil trace. Obviously also the longer a species survived, the greater the chances of fossil remains being found. During periods of speciation it is reasonable to assume that some 'transitional' species were around only briefly, thus reducing their chances of leaving fossil remains.

This does not answer the 'no transitional fossils' argument completely, but perhaps helps to put it into perspective. But here we encounter a rather odd situation. When such remains *are* found, Creationists denounce them as fakes and refuse to accept them (a notorious example is *Archaeopteryx*, which is transitional between dinosaur and avian species). And where, as with the horse lineage, there is a fairly convincing sequence of fossils, the response is often simple denial that they represent what they appear to represent.[3] Actually there are many continuous fossil lineages in which the transition of one species into another can be traced, especially among marine molluscs. In these cases quite opposite difficulties can occur. The problem becomes a methodological and theoretical one for taxonomists, precisely because the exact criteria for differentiating species can become unclear (interfertility cannot be used with fossils!). Even though everyone agrees that speciation has occurred over time, the earliest and last specimens being quite distinct, the precise point at which it occurred becomes impossible to pin down 'objectively'. Two examples are the Liassic oyster genera *Gryphaea* and the Carboniferous coral *Zaphrentis delanouei*. These were studied as early as 1924 by A.E. Trueman and in 1910 by R.G. Carruthers respectively. The term 'chronological subspecies' was used to describe the successive forms of these examples.[4]

The volume of human and hominid fossil evidence continues to grow and is now far greater than it was even 30 years ago, but its geographical and temporal distribution is still extremely patchy. While new hominid fossils can still force palæontologists to rethink their current, tentative, hypotheses about the course of events, they have found no reason to rethink the fundamental picture of a relatively gradual evolution from a common human–chimpanzee ancestor to modern humans. Whether a find is 'transitional' or a blind alley side-branch is often uncertain, but each new fossil helps to fill out the jigsaw in more detail, and none appears to come from a different jigsaw altogether.

No honest evolutionary biologist would claim that the speciation process is entirely understood as yet, and various models are in play, but these are the kinds of problem that any actively progressing science faces. The 'fossil gaps' argument was abandoned by anti-evolutionists for quite a long period because, as mentioned in the main text, it was an obvious hostage to fortune, and periodically transitional fossils *do* come to light. Its resurrection among American Creationists only reinforces points made in Chapter 5 about the sociological nature of American Creationism. If as many resources were allocated to discovering transitional fossils as are devoted to discovering oil, I am sure that progress on the matter would accelerate rapidly. (The ID camp has switched its focus to finding as yet unsolved problems in what are actually new and rapidly developing research areas, typically in microbiology and genetics – precisely where one would expect to find them of course, almost by definition.)

The 'incredulity' factor

Lay people often find the completeness of fossils extremely hard to reconcile with the great ages that geologists ascribe to them. How can they possibly have survived in such perfect condition for tens of millions of years? But if the circumstances for fossilization are rarely favourable, on occasion they can be very favourable indeed. We have to remember the extreme slowness of many processes over geological time-scales. One location, where tectonic plates collide, may be constantly ravaged by volcanoes, earthquakes, and pyroclastic flows, but another may remain undisturbed by anything other than slow climate changes or sea-level shifts. Take the example of a shallow seashore lagoon, flooded at every high tide. As the tide recedes, a large number of fish and crustaceans are temporarily cut off from the sea. But if the land is very slowly rising, or the sea level falling, there comes a day when the rising tide does not quite reach the lagoon, and will never do so again. The organisms are trapped there for ever. Their bodies gently fall to the bottom of the lagoon. There are no scavengers to eat them. The chemical conditions are not corrosive. The wind blows sand into the lagoon, which gradually dries out, covering the undisturbed remains with fine-grained sediment. No longer washed away by the sea, the layers build up and consolidate around the remains, preventing them from being squashed by the weight of any overlying deposits. After that, *nothing can happen to them* until they are either exposed by later erosion, destroyed by an intrusion of volcanic magma, or dug up by the fossil hunter (I do not claim that this list is entirely complete!). So why would they not be as pristine as the day they died? There are other scenarios, such as being trapped in a cave by a rock-fall and falling into a tar-pit, as well as less sensational circumstances, which can easily be imagined as resulting in prolonged, undamaged, internment. One must also stress that fossilization proper involves a slow and complex chemical replacement of the original organism's material by other minerals, preserving its form but not its actual substance. If you doubt all this, then you are advised to study the sciences involved. Incredulity is not evidence. Isn't it obvious that the Sun goes round the Earth?[5]

The fossil scale

The opponents of evolution are forever looking for cases where a fossil does not occur where it 'should' do according to evolutionary palæontology. There are two simple things to be said about this. First, that nobody is claiming to have a complete picture of the evolution of life on Earth, so every so often it *is* likely that a fossil specimen of a species will be found outside the expected time-frame, especially in regions previously unexplored. That is part of the reason why fossil hunting is so appealing. Secondly, the fossil scale is always provisional and, to repeat an earlier metaphor, we are dealing with a jigsaw of a picture of which we have still only a rather blurred image, and of which the overwhelming majority of the pieces are irrevocably lost. What is impressive is not the odd anomalous or baffling fossil, but the extent to which additional fossils are constantly filling out the picture more or less along the lines that we expected on the basis of evolutionary assumptions. The more extreme examples often cited, such as fossilized human footprints alongside those of dinosaurs, have all, to my knowledge, been refuted as products of misinterpretation or fraudulence. The most famous of these cases, that of the 'Paluxy' tracks, found in Texas, which allegedly showed human footprints alongside those of dinosaurs, is now thoroughly discredited, as most Creationists have accepted.[6]

Fossils and the Flood

A few Creationist websites, such as the second mentioned in Chapter 1, Note 1, continue to insist that fossils are remains of animals drowned in Noah's Flood (even, I assume, fossil fish). If you seriously think this is a possibility, you are either beyond the scope of reasonable debate or ignorant of the entire subject and need to go away and do some basic research. Visit Lyme Regis maybe.[7]

To end this brief excursion into a very specialized topic, I will simply say that I have seen nothing in the Strong Creationist, 'Young Earth' critiques of fossil evidence that remotely challenges even my own, relatively rudimentary, grasp of the subject.

Appendix B: Pondering on *Genesis*

Leaving aside current 'scientific' ID theorizing, it is useful to ask what we are to understand is being proposed by literal 'Six Day' 'Young Earth' Creationism. Let us take a rather dead-pan approach, pretending that we are reading this account for the first time, using the Revized King James Version of the Bible, without any prior knowledge of it. *Genesis* 1–7 paints a rather puzzling sequence of events. First God creates the world in six days, finally fashioning Adam to share the experience and take care of the Garden of Eden; then, seeing that it is good, God takes a day off. So what was this 'Prelapsarian' world like? This is where I encounter my first problem. It certainly contained Ethiopia, Assyria, and the river Euphrates (*Genesis* 2:13–14). But how far did the Earth's geography otherwise resemble that which it has now? Nobody disputes that even over historical time cliffs have receded, volcanoes and earthquakes have transformed landscapes, sea-levels have risen and fallen, rivers changed their courses, coastlines variously grown and eroded, and so forth. There would, I assume, have been no fossils – and hence no chalk or coal strata – since literal Creationists argue that they are either remains of creatures killed in the Flood or (although this is argued more rarely nowadays) that they were placed there by God to baffle and mislead Fallen man by way of testing his faith. Surely God would not have created them before the Fall as a kind of practical joke on His best (and presumably only) friends Adam and Eve? It is also unclear whether in this initial paradise there was any death, but at any rate there were no thorns and thistles, because these are not created until *Genesis* 3:18 as part of Adam and Eve's punishment. (And what previously sustained those creatures that live on them?) As mentioned in the main text, the Earth's current imperfections are typically explained as due to our living in a Fallen world. I suspect that the Creationists' image is of an initially perfect, unchanging, harmonious world – the kind of 'peaceable kingdom' at one time depicted by artists in which the lion lay down with the lamb. But why would God have designed predator species so well – only to inhibit them from exercising their intended life-style?

When Adam and Eve 'fell', they must, on this view, have taken everything else with them and thus inaugurated the cycles of death and suffering which all of animate nature now endures. Presumably too there could previously have been no destructive floods, earthquakes, and volcanoes (let alone meteor impacts). Were the cliff-faces all shiny and smooth? Did the tides ebb and flow in gentle ripples? And did continental India suddenly rush across the Indian Ocean to hit Asia and create the Himalayas as the flaming sword began to twirl at the gates of Eden? Was it now that God also added confusion for Adam's heirs by inserting innumerable folds, buckles, faults, and igneous intrusions (but not yet fossils perhaps) into a hitherto neat geological layer-cake? I suppose there must have been some hills before the Fall, otherwise the rivers could not have flowed, but huge mountain ranges and deserts with their constant landslides, glaciers, and choking sandstorms?

What this implies is that there was, so to speak, a second episode, 'Creation II', accompanying the Fall. This raises a serious question of how far the initial creation is supposed to have resembled the world we now know, and how far it was over-ridden by this sequel. But the series is not over, for next we have Noah's Flood, which again wipes the slate clean as far as 'every living substance ... upon the face of the ground ... and the fowl of the heaven' were concerned (*Genesis* 7:21–23). Noah himself now has to embark on 'Creation III' by breeding from the animals that he had taken on board his ark. It is not explained how the herbivores managed in an initially vegetation-free world, and no mention is made in *Genesis* 7 of his taking a comprehensive seed-bank on board to replenish the plants that had been wiped out. (He would have needed thorn and thistle seeds for the donkeys.)

Many modern Creationists also continue to hold that the Flood itself was responsible for fossils, for the Grand Canyon, and, one has to assume, for all rocks and minerals of organic origin, such as coal, oil, flints, and chalk. Quite how, in the few weeks that it lasted, the Flood could have laid down the numerous vastly extensive fossiliferous strata through which it then carved the Grand Canyon (and presumably Bryce and Zion Canyons too) as it receded is something of an additional conundrum. Why so many fossils are of marine creatures is even more baffling, because these are not mentioned in *Genesis* 7 as being among those destroyed, and the Flood would not have caused them much inconvenience at all – or perhaps the seas became temporarily brackish? But there are fresh-water fossils too. If the event was momentous enough to create the Grand Canyon, its effects on the landscape must have been dramatic indeed – but, oddly, not enough to change the course of the Euphrates, which, as noted above, was there from the start.

Far from being simple, the account of the origins of the present world as provided in the opening chapters of *Genesis* raises far more questions than it answers regarding how we can possibly imagine the relationship between the three episodes of initial cosmic Creation, Fall, and Flood. And fitting them all into the last 6,000-odd years is an equally puzzling challenge, especially given that the Flood did not take place until a couple of thousand years or so after the Fall, if we are to believe the life-spans given in *Genesis* 5, and that Noah was already 600 years old when it happened, five years after his father Lamech had died (*Genesis* 7:6).

In truth, of course, biblical scholars have, ever since Richard Simon's epochal 1678 *Histoire Critique du Vieux Testament*, appreciated that the Old Testament is a complex multi-authored text with a chequered history of rewritings, censorship, and re-editing. Most scholars also now acknowledge that the early chapters of *Genesis* retain hints of a pre-monotheistic origin, and that the serpent in the Garden of Eden can hardly be equated to Satan – we are only told 'Now the serpent was more subtil than any beast of the field which the Lord God had made' (*Genesis* 3:1), without any further theological exegesis whatsoever. Even so, it is odd to describe a fallen Angel as a 'beast of the field'. There is also an inconsistency between *Genesis* 1:26–27, in which God 'created man in his own image ... male and female he created *them*' (my italics), and *Genesis* 2: 8–18, in which He creates a single Adam and then, from one of Adam's ribs, his female partner, whom Adam does not actually call Eve until after the Fall (*Genesis* 3:20) 'because she was the mother of all living', as if she was a mother goddess (the Hebrew version *Hawwah* meaning 'life'), which, again, is somewhat strange, given that we have previously been told that God created all living things.

It is surely only long and constantly repeated uncritically reverent familiarity with these chapters that has given them a patina of overall coherence for lifelong-devout Christians and Jews. Coming to them 'cold', as it were, and rereading them afresh for the first time in decades, I am confronted with unmistakable hallmarks of ancient re-editing and multiple origins.

The relationships between the *Genesis* account and other creation myths from ancient Canaanite, Sumerian, and Babylonian cultures have been examined at great length since the mid-nineteenth century. The chapter on Hebrew mythology in *Middle Eastern Mythology* (S.H. Hooke, 1963) is an excellent introduction to the whole issue, on which I am drawing in what follows, although biblical scholarship has obviously moved on hugely since this was published.[1] It is now generally accepted that the first two chapters of *Genesis* contain two distinct creation myths. The first is the 'Priestly' myth, now thought to have been written later than the second, being produced by 'the priestly editors who collected and arranged the traditions of Israel after the exile' (Hooke, p.109). This is rooted in the Mesopotamian Sumerian and Babylonian creation myth which describes the original state of the universe as a watery chaos. The second myth, 'Yahwist' and 'Elohist', dates from the early period of the kings of Israel as part of an attempt to bring together ancient oral and written traditions into a single account. That this involved two schools or writers is signified by the uses of both *Yahweh* and *Eloi* to refer to God. This version is rooted in a very different Canaanite myth, in which the universe starts as a waterless waste.

The Flood story is now widely assumed to represent a folk-memory of the widespread inundations which would have accompanied the end of the last Ice Age around 10–12,000 years ago. Hooke presents a two-page table detailing the parallels between the Sumerian and Babylonian flood myths and the two (intertwined but distinct) biblical Yahwist and Priestly versions. And so it goes on. Readers might wish to consult http://mb-soft. com/believe/txs/genesis.htm for a detailed analysis of the intertwining of the various authorships of the first 10 chapters of *Genesis*. This is a non-Creationist Christian website, but, although summarizing the history of authorship analysis from the eighteenth century onwards, it does not explore relationships with other mythological traditions. Creationists are, of course, well aware of the issue and are striving to contest multiple authorship on various grounds. A Google search in August 2006 for Genesis+authors OR authorship yielded 9.28 million hits, which, even allowing for mis-hits such as a publishing company called Genesis and sites dedicated to the pop-group Genesis, demonstrates how heated the matter is. In fact, a less-noticed battle parallel to that which we have been addressing in this book is also being fought between Creationists and academic scholars on this front.

In conclusion, it is not that I do not believe the creation story as given in *Genesis*: I do not even understand what it is I am being asked to believe by those who say they do believe it. And the more I ponder on it, the less coherent it becomes. But then perhaps some Creationists feel that we are just not supposed to ask such questions, and that takes us even further from the realms of scientific enquiry. Curiously, one might note that, from the Fall onwards, the God of the Old Testament gets particularly annoyed by human attempts to gain knowledge, which He feels to be assaults on His own omniscience, *Job* 38–39 being a famous case in point. So, in siding with Him on this, such Creationists would at least be being consistent with their faith.

Appendix C: *Mythos* and *Logos*[I]

In the main text I have referred on several occasions to the distinction between *mythos* and *logos* which Karen Armstrong has deployed in her recent works. While I did not wish to frame my entire work in terms of this distinction, it does bear so closely on the central issues that some further exposition of it is required. Although the distinction has a long history, it had not, I think, prior to Armstrong's work been widely used, except in some theological contexts. There are some more familiar near-equivalents: arts versus science, narrativism versus logic, imagination versus reason, and the like. None of these, however, has an equivalent generality, and they may be seen as facets of this more fundamental distinction. As I read it, the distinction may best be understood as differentiating between two absolutely fundamental and complementary psychological functions necessary for civilized human life at both individual and socio-cultural levels.

Mythos

To take *mythos* first. This refers to the necessity for some kind of over-arching framework or system of values and meanings in terms of which our lives, both socially and personally, are given some kind of significance. This includes our relationship to, and place within, the cosmos at large, as well as our personal biographies and experiences. A *mythos* is not a set of empirical knowledge claims or a methodology for acquiring empirical knowledge, nor a set of rules for governing political and economic policies. But it is only *within* a *mythos* that such activities can occur in the first place. It is *mythos* which defines their meanings for the societies in which they happen. As we have seen, for example, the *mythos*-type notion that scientific knowledge was an exploration of a divinely created 'Book of Nature' provided the very rationale for its pursuit until the latter years of the eighteenth century. Equally, economic doctrines such as the virtue of the 'Free Market', or indeed Socialism, are underpinned by *mythos*-provided values of various kinds. Traditional religions and ancient mythologies are obviously the most fully articulated and comprehensive expressions of the *mythos* function, but they do not exhaust it. It is only within a *mythos* that we can construe the events of our lived experience in a fully meaningful way. A *mythos* is expressed or communicated in many forms, such as moral rules (the Ten Commandments), symbols (the Christian Cross and the Ka'aba in Mecca), abstract concepts (Equality, Free Will), and of course in art, poetry, drama, literature, and music.

In serving this function, such rules, symbols, concepts, and creative expressions can acquire enduring value when they pertain not to specific historical circumstances but to concerns which are virtually universal: love, death, justice, human relationships, beauty, emotion, and our place in the universe. They are about how to live a fulfilled life and handle its failures to be fulfilling – about the wisest ways to respond to joy and disaster.

Logos

Logos, on the other hand, is about practical problem solving, identifying in a pragmatic but logical fashion the ways in which the physical and social worlds operate. *Logos* can help you to win an election and enable you to build a new weapon, explore the stars, travel from Los Angeles to Baku, solve or commit a crime, and a million other things. It can also help you to understand human nature itself, as the study of Psychology tries to do, but there are limitations as to how far it can go with this and still remain *logos*. Sometimes *logos* reveals truths as enduring as the products of *mythos* – the Earth goes round the Sun, for example, or $E=MC^2$ – but more typically its products are transient. Its conclusions are shown to be wrong, the problems concerning it disappear, and the technologies that it yields become obsolete. If enveloped in *mythos*, *logos* is required to apply its values and principles in everyday life. The ruler who disregards *logos*, believing himself or herself to have a direct line to the deity of their *mythos* and thus be infallible, is heading for disaster. But the ruler who disregards *mythos* rapidly becomes morally rudderless and equally disconnected from reality; while, at some level, guided by *mythos*, *logos* can, with equanimity, over-ride its injunctions if the circumstances demand it. 'Thou shalt not kill' – but capital punishment and wars continue; 'Love thine enemies' – but bomb the hell out of the Taliban; 'Allah is merciful' – but behead your hostages. This is often referred to as a failure to 'live up to our ideals', or a rueful acceptance that these 'ideals' are in practice 'unrealistic'. But this does not mean that they should nihilistically be abandoned. That this bears on the deep US cultural tension referred to earlier should be obvious.

Since both *mythos* and *logos* are inherent aspects of psychological functioning, negotiating their relationship without allowing either to gain supremacy has been a perennial task. (Roman Catholic Popes long ago mastered this art.) That we try implies, however, that there must be a third 'synthetic' function which is also psychologically essential.

And that brings us back to Psychology. As a would-be natural science it has perennially cast itself, for the most part, in the *logos* camp. It has frequently evaded confronting moral issues by invoking the 'facts versus values' distinction, although the neatness of this is now generally acknowledged to be illusory. But directly concerning human nature and experience as it does, its popular appeal has actually rested on the assumption that it *can* illuminate matters of value, morality, and personal fulfilment, the territory of *mythos*. There is, however, a further vicious twist. Is the *mythos* versus *logos* distinction in itself a *logos*-type, 'Psychological', empirical hypothesis? Or is it a *mythos*-type response aimed simply at providing a meaningful image of the situation that we now face? The reconciliation and integration of *mythos* and *logos* has traditionally been managed by claiming that the prevailing *mythos* has been ratified or confirmed by empirical facts – the historicity of Christ's passion being an obvious example. Clearly this solution can no longer carry the day. What the Creationism/ID controversy brings home is the urgency of the task of conceptualizing the *mythos–logos* relationship afresh in a way that is satisfactory to religious and scientific parties alike – a conceptualization, moreover, which derives its authority from neither.

At this point perhaps we need a philosopher (a 'lover of wisdom'). It would be wisest of me to leave it there, I think.

Notes

Chapter 1

1   The leading Creationist website is http://www.icr.org, posted by the Institute of Creation Research. More naïve and extreme is www.creationism.org, whose arguments concerning fossils can only be described as ludicrous. (There are many other websites supporting Creationism and/or ID, but few, whatever my issues with them, would I describe as 'ludicrous'.)

2   K. Armstrong (2000) *The Battle for God. Fundamentalism in Christianity, Judaism and Islam*. But see also Paul Tillich (1952), to be discussed in Chapter 9, for a more profound existentialist-cum-psychological analysis.

3   A very substantial academic literature has developed over the last half-century, examining the topic from a variety of historical and philosophical perspectives. In recent years one prolific contributor has been Michael Ruse, beginning with his 1988 edited work *But Is It Science? The Philosophical Question in the Creation/Evolution Controversy*. His most recent work is *Darwinism and Its Discontents*. Another major authority is Ronald L. Numbers, whose most important work on the topic is *The Creationists: The Evolution of Scientific Creationism* (2006). More generally, see Robert T. Pennock (ed.) *Intelligent Design, Creationism and Its Critics: Philosophical, Theological, and Scientific Perspectives*, and Nathaniel C. Comfort (ed.) (2007) *The Panda's Black Box: Opening up the Intelligent Design Controversy*, which includes a chapter by Ruse and one by R.M. Young which is particularly useful. One succinct recent critique is the late Walter M. Fitch's *The Three Failures of Creationism. Logic, Rhetoric, and Science* (2012). Written by an evolutionary molecular biologist, this is somewhat different in approach from the present work, focusing more narrowly, but in greater depth, on specific scientific arguments and issues, while adopting a more directly pedagogic tone. I feel that our texts should be considered as complementary. Another critique from a detailed scientific perspective is Sahotra Sarkar (2007) *Doubting Darwin? Creationist Designs on Evolution*, which is extremely useful.

4   See Karen Armstrong's article 'Bush's fondness for fundamentalism is courting disaster at home and abroad', *The Guardian*, 31 July 2006, p.24, for a disturbing overview of this.

5   See Ronald L. Numbers (1987) 'The Creationists' (also his 2006 *The Creationists*, previously cited).

6   See website: freemasonry.bcy.ca.

7   There is a Wikipedia entry on 'Ussher chronology' which looks fairly reliable.

8   On the anti-Newtonian resistance of the 'mechanists', see E.J. Aiton (1972).

9   A useful review of this topic is F.C. Haber (1959) *The Age of the World: Moses to Darwin*. There is probably also a more recent treatment of which I am unaware.

10 This was built on sixteenth- and seventeenth-century work by humanist scholars such as Erasmus, culminating in Richard Simon's *Histoire Critique Du Vieux Testament* of 1678, which concentrated on establishing reliable biblical texts. The later German school of historical, as opposed to textual, criticism is associated with H.S. Reimarus, G.E. Lessing, W. Gesenius and others. A parallel but distinct development was the liberal Protestant 'hermeneutic' school associated with F. Schleiermacher.

11 I cannot venture into the entire, complex topic of the history of religion in the United States. Two relevant works are Nathan Hatch (1989) *The Democratization of American Christianity*, and Mark Noll (2002) *America's God: from Jonathan Edwards to Abraham Lincoln*.

12 As an aside here, I might note that in 1902 one devout writer proposed an argument to salvage the *Genesis* story which seems to have been forgotten but is worth noting. This is that it was on these days that God created the laws governing the phenomena in question, which then played themselves out over an unspecified subsequent time period to bring the physical creation itself to fruition (F.H. Capron, 1902, *The Conflict of Truth*). Ingenious, if nothing else.

13 For a thorough overview of American fundamentalism during the twentieth century, to which much of the subsequent coverage of this topic is indebted, see Karen Armstrong (2000) *op. cit*, n.2, Chapters 6–10. This is an excellent and learned book, strongly recommended as general background reading. Also to be recommended for its coverage of the educational aspect of the American story is Dorothy Nelkin (1982) *The Creation Controversy. Science or Scripture in the Schools*.

14 Two relevant works on this by Edward J. Larson are *Trial and Error: The American Controversy over Creation and Evolution* (1987) and *Summer for the Gods: The Scopes Trial and America's Continuing Debate over Science and Religion* (1998).

15 All were published by the Pacific Press Publishing Association, Mountain View, California. For some time Price as a Seventh Day Adventist was held at arm's length by early twentieth-century American Fundamentalists, although he rose in status between the world wars.

16 See Ronald L. Numbers (1987) *op. cit.* for a fuller account of the groups and individuals involved in mid-twentieth-century American Creationism.

17 Chapters 9–10 of Armstrong (2000), cited above, n.2, treat this in considerable detail.

18 See http://www.talkorigins.org/indexcc/CB/CB200_1.html for rebuttals of this, including the fact that there are indeed organisms with simpler versions than that discussed by Behe. This website should be considered as a first port of call for those seeking answers to specific Creationist and ID arguments, particularly empirical 'scientific' claims, if it is not suffering from one of the periodic cyber-attacks

that it predictably attracts. See also www.millerandlevine.com/km/evol/design2/
article.html for an accessible overview of mainstream scientific rebuttals of Behe's
argument. An enormous amount of effort has in fact been expended by numerous
biologists and other scientists in refuting Behe. Sahotra Sarkar (2007) *Doubting
Darwin* (see Chapter 4, note 3) is especially thorough. But I cannot help asking: what
kind of omnipotent cosmic Designer would be so strangely coy about their prowess
as to bury the only unambiguous piece of evidence for it at the level of the flagellae
of a microscopic bacterium? The genuinely earnest passion and straightforward
bewildered anger of Behe's scientific opponents has perhaps blinded them to just
how ludicrous his position really is.

19   The official Catholic position on Creationism and ID is somewhat cautious:
     although it clearly has little time for Young Earth versions based on 'literal' readings
     of *Genesis*, it also tends to accept evolution in all cases except that of humans. The
     'creative' role in the human case, however, was the 'infusion' of a soul into a pre-
     existing human-like species. When the chips are down, Catholic experts have sided
     with anti-Creationists in US legal cases. There are numerous websites relevant to
     this (search Catholicism + Creationism), but see particularly www.americancatholic.
     org/Newsletters/CU/ac1007.asp, which is written by Michael D. Guinan.

Chapter 2

1   James Martineau (1805–1900) was a typically prolific Victorian author of a number
    of multi-volume works and arguably the century's most eminent Unitarian thinker.
    His writing style is actually rather good, and on the topic in question the most
    accessible introduction to his position is probably the 68-page *Religion as Affected
    by Modern Materialism. An Address delivered in Manchester New College, London*. This
    lecture was given in October 1874 and was published the following year by Putnams
    in New York. See also his late *The Seat of Authority in Religion* (1890).

2   Other than modern humans, of course, who certainly do 'design' their appearance
    and can even, using implanted chips, perhaps redesign aspects of their physiological
    functioning. Genetic engineering of babies does not count, because it is not the
    baby who 'designed' itself.

3   How the Fall would apply to the extraordinary course taken by the mammalian
    laryngeal nerve is quite unclear. Running from the brain to the larynx, a distance of
    only a few inches, it runs all the way down to the chest, loops round a blood vessel,
    and returns all the way up again, even in the giraffe. Evolutionists can explain this
    quite simply: it first appears as a nerve in primitive neckless fishes, linking the brain
    to the gills, and gradually, as necks extended etc., it became progressively longer.

4   This was long ago well reviewed in W.M. Krogman (1951) 'Scars of evolution', in
    Scientific American, and later touched on by C.O. Lovejoy (1981) 'The Origin of Man'
    in the leading US science journal *Science*.

5   '... there must have existed, at some time, and at some place or other, an artificer *or artificers*, who formed it for the purpose we find it actually to answer' (W. Paley *Natural Theology*, Chapter 1, my italics).

6   For a survey of the way in which Nature has been conceptualized and understood in Western cultures since antiquity, see Peter Coates (1998) *Nature. Western Attitudes since Ancient Times.* He does not address the Creationism issue directly, but his coverage of Christian attitudes clearly has bearings on it.

Chapter 3

1   His highly successful and popular 1926 *Science and the Modern World*, later reissued as a Pelican paperback, is a good introduction to his thought.

2   G. Richards (2005) 'The Psychology of Explanation'.

3   Reported on http://www.worldnetdaily.com/news/article.asp? ARTICLE_ID=51128

4   See C.G. Jung's 1955 essay 'Synchronicity. An Acausal Connecting Principle' for a particularly curious discussion of this. (This is in C.G. Jung and W. Pauli *The Interpretation of Nature and the Psyche*, along with an essay by Pauli not directly relevant here.)

Chapter 4

1   See B. Latour (1987) *Science in Action. How to Follow Scientists and Engineers through Society.*

2   Hilary and Steven Rose (eds.) (2000) *Alas, Poor Darwin. Arguments Against Evolutionary Psychology*, pp. 85–105.

3   For an accessible, but rather sensationalized, summary of HGT, see Graham Lawton (2009) 'Uprooting Darwin's tree', *New Scientist*, 24 January. This issue must have caused Creationists' hearts to skip a beat, as the cover proclaimed 'Darwin was wrong. Cutting down the tree of life'. To discover that what he was wrong about was the tree image, that it was not being felled but converted into a web, and that the new developments reported were reinvigorating evolutionary theory, rather than threatening it, must have been something of a let-down. For an excellent fairly recent view of the nature of evolutionary theory by an eminent molecular biologist-cum-philosopher, see Chapters 4 and 7 of Sahotra Sarkar's (2007) *Doubting Darwin? Creationist Designs on Evolution.*

4   There are in fact a whole range of dating methods based on a variety of disciplines and different kinds of data. What is remarkable is the degree of consistency between the results from these radically different approaches.

5   Readers wishing to explore degenerationism and related topics should start with Daniel Pick (1989) *Faces of Degeneration. A European Disorder c.1848–c.1919.* On

eugenics the best introduction, if now a little out-dated, remains D.J. Kevles (1985) *In the Name of Eugenics*. On scientific racism a major early historical review is J.S. Haller Jr (1971) *Outcasts from Evolution: Scientific Attitudes of Racial Inferiority 1859–1900*, while I had my own stab at it from the point of view of its influence within Psychology in G. Richards (1997, 2ⁿᵈ ed. 2012) *'Race', Racism and Psychology. Towards a Reflexive History*. There are, however, quite extensive historical literatures on these overlapping topics. For a variety of criticisms of the continued use of evolutionary concepts in 'evolutionary psychology' and some sociology, see Hilary and Steven Rose (eds.) (2000) *Alas, Poor Darwin. Arguments Against Evolutionary Psychology*, the papers by Mary Midgley, Tim Ingold, and Steven Rose being particularly salutary.

6 W.M. Fitch (2012) *The Three Failures of Creationism. Logic, Rhetoric, and Science* (pp.136–7) argues that while stated in these terms the proposition is indeed tautological, and it does not actually correctly represent the evolutionary position: 'Those that survive are the fittest' should be replaced with 'The more fit are those better adapted to their environment', which introduces the non-tautological principle of adaptation. This is perhaps fair enough, but it does not affect my own argument here about the implications of tautology.

7 Morris's position, initially published in a June 2000 pamphlet entitled 'The Vital Importance of Believing in a Recent Creation', is posted on the web at www.icr.org/pubs/btg-a/btg-138a.htm.

Chapter 5

1 For an introduction to the various ways in which the Bible has been read, K. Armstrong (2007) *The Bible: a Biography* is extremely useful. Drastically simplified, there appear to be four major approaches: (a) although the Bible is the revealed 'Word of God', its contents are of different kinds: literal facts, allegories, prophecies, moral teachings, etc., and the reader's task is to interpret which of these is most appropriate (though accepting that a passage may fuse more than one, particularly in the Old Testament): this is perhaps the most 'orthodox' approach; (b) the Bible is a route for communicating with God, it has infinite potential meanings, none is final, and more immediately the task is to understand God's will at the present time: this is the Jewish *Midrash* tradition, initiated with the destruction of the Temple in Jerusalem, and at its crudest it can almost take the form of fortune-telling: the Bible is opened at random and a finger blindly directed to a verse which, it is believed, will give 'guidance' (Armstrong does not discuss this); (c) the scholarly tradition exemplified in the late eighteenth-century emergence of the 'Higher Criticism', in which the aim is to recover the original meanings and contexts of biblical texts, how they were composed and subsequently re-edited, etc.; and (d) the modern 'literalist' tradition. Incidentally, the eminent Methodist minister and psychotherapist Leslie D. Weatherhead has a short chapter on 'The Bible and its inspiration' in his 1965 *The Christian Agnostic*, which is refreshingly frank about what he sees as its highly variable, and sometimes quite objectionable, character.

2 Technically, Scofield's position is described as 'dispensational premillenialism'. A full, but sympathetic, book-length study of the place of the Scofield Bible in US Christianity has now been issued: R. Todd Mangum and Mark S. Sweetnam, *The Scofield Bible: Its History and Impact on the Evangelical Church*. Scofield's dates were 1843–1923. The Wikipedia entry seems acceptable, although I cannot vouch for its complete accuracy. Scofield also published a multi-volume *Bible Correspondence Course* and leaflets on the topic. I used the 1917 'New and improved' edition (published by Oxford University Press). I infer from the Wikipedia entry that a Rev. William L. Pettingill must have been added to the associate editors since the first edition.

Chapter 6

1 H. Zimmer, ed. J. Campbell (1946), *Myths and Symbols in Indian Art and Civilization* – much reprinted – is my source for this, but undoubtedly there are many others.

2 This is cited in W. Elfe Turner (1855) p.260. Turner's book contains an interesting marshalling of the evidence against extended geological time-scales in order to demonstrate the unreliability of geologists' figures of these. There is already, however, something of a 'last ditch' tone about it.

3 A simple Google search for 'Flat Earth' will easily enable the reader to enter the Flat Earthist world, but it should be noted that since Johnson's death the groups and individuals claiming to believe the doctrine appear to be mostly satirical or at best tongue-in-cheek. Since the very satellite communications technology on which the Internet relies is itself premised on the Earth's being globular, it is, of course, hard to see how the doctrine can any longer be advocated with anything remotely resembling seriousness.

4 It is nice to be able to cite *The Big Issue*. This is quoted in 'Border control', an interview by Stephen Appelbaum with film director Tommy Lee Jones, *The Big Issue*, No. 686, March 27–April 2 2006, p.15.

Chapter 7

1 There are a few examples of this, starting perhaps with Auguste Comte, and continuing with some evolutionists such as Francis Galton and Raymond B. Cattell. They have generally ended up looking faintly ridiculous.

2 I am using the term 'philosophy of life' in a loose sense here to cover all broad frameworks of meaning in terms of which we live our lives; these include religion, but of course religion is not to be confused with philosophy in the strict sense as an academic discipline – or even indeed with theology!

3 G. Richards (2010 3rd ed.) *Putting Psychology in its Place. A Critical Historical Overview*.

4   Two further recent works of Roger Smith should also be noted as shedding important light on the nature of Psychology and how it has engaged the MBP: *Free Will and the Human Sciences in Britain, 1870–1910* and *Between Mind and Nature. A History of Psychology* (both 2013).

5   For a very subtle recent account of the history of the paranormal and spiritualism since the early 1800s, see P. Lamont (2013).

6   Two useful twentieth-century treatments are G. Dawes Hicks (1937) *The Philosophical Bases of Theism* and A. Boyce Gibson (1970) *Theism and Empiricism*. The preceding Martineau quote is taken from W.B. Carpenter (1888) *Nature and Man. Essays Scientific and Philosophical*, which does not supply its precise source.

7   I have elaborated on this in G. Richards (1989) *On Psychological Language and the Physiomorphic Basis of Human Nature*. Even the great physicist Erwin Schrödinger confessed himself baffled by this in his 1956 lecture *Mind and Matter* (Schrödinger, 1958).

8   Physicists would, rightly, object that space and time *do* have certain properties of their own, but these relate to the physical world in a way crudely analogous to the way in which the properties of a film screen relate to the films being shown. They have effects on the film and are necessary for the film to 'happen', but are not in themselves part of the film.

9   Otto's classic exposition of this was *The Idea of the Holy*, first published in German as *Das Heilige* in 1917 and in English in 1923, going through several subsequent editions.

10   See G. Richards (2011) *Psychology, Religion, and the Nature of the Soul. A Historical Entanglement* for the author's fuller view of this topic.

Chapter 8

1   There is a huge academic literature on the science–religion relationship, written from many historical, philosophical, and religious perspectives and extending back to the mid-nineteenth century. A major early work was J.W. Draper (1875) *History of the Conflict between Science and Religion*. In the last century it was a perennial theme of popular science writings by people such as Joseph Needham and James Jeans, and more recently by the philosopher Mary Midgley and numerous others. The Templeton Foundation is a major contemporary source of funding and support for work striving to bridge the divide. The present chapter is but a rapid sweep through the topic. A short recent (and widely praised) overview of the science–religion relationship is Thomas Dixon (2008) *Science and Religion: A Very Short Introduction*, which could well be taken as a starting point.

2   D. Nelkin, 'Less selfish than sacred? Genes and the religious impulse in evolutionary psychology' in Hilary and Steven Rose (eds.) (2000) *Alas, Poor Darwin. Arguments Against Evolutionary Psychology*, pp.14–27.

3   James E. Lovelock (1979) *Gaia: A New Look at Life on Earth.*

4   This phrase entered circulation after the 1874 publication of J.W. Draper's *History of the Conflict between Religion and Science.* His main target, however, was Roman Catholicism's dealing with science, not the Argument from Design.

5   Steve Fuller (2008) *Dissent over Descent. Intelligent Design's Challenge to Darwinism.* For a somewhat bad-tempered exchange between the author and Steve Fuller, see my 2009 review of the book in *History of the Human Science,* with Fuller's response and my reply to this.

6   James Jeans (1930) *The Mysterious Universe*; Alister Hardy (1965, 1966) *The Living Stream, The Divine Flame* (these were his Gifford Lectures); Fritjof Capra (1972) *The Tao of Physics. An Exploration of the Parallels between Modern Physics and Eastern Mysticism*; and the works of Teilhard de Chardin, the most popularly successful being *The Phenomenon of Man* (1959).

7   See G. Richards (2011) Chapter 3 for a brief consideration of the topic of Catholics in Psychology, with references to more substantial works. R. Kugelmann (2011) is the most thorough and comprehensive coverage.

Chapter 9

1   See Chris Mooney (2005) *The Republican War on Science* for a general exposé of this entire topic, including a short chapter entitled '"Creation Science" and Reagan's "Dream"'.

2   William H. Tucker (1994) *The Science and Politics of Racial Research.* The subsequent literature on this topic by Tucker himself and others is becoming progressively even weightier.

3   See also M. Midgley (2007) *Intelligent Design Theory and Other Ideological Problems.*

Appendix A

1   The most recent comprehensive work on the topic appears to be J.-P. Cuif, Y. Dauphin, and J.E. Sorauf (2010) *Biominerals and Fossils through Time.*

2   Pat Shipman (1981, reprinted 1993) *Life History of a Fossil and Introduction to Paleoecology.* A recent publication on taphonomy is T.R. Pickering (ed.) (2007) *Breathing Life into Fossils.*

3   On the evolution of the horse, see B.J. MacFadden (1992) *Fossil Horses. Systematics, Paleobiology and Evolution of the Family Equidae.*

4   For discussion of these examples, see Norman D. Newell 'Fossil populations' in P.C. Sylvester-Bradley (ed.) (1956) *The Species Concept in Palaeontology.*

5   Although, as Wittgenstein allegedly once asked, when someone said that
    heliocentric beliefs are easily understandable because it does look as if the Sun goes
    round the Earth, 'What should it look like then?'

6   For a good overview, see http://paleo.cc/paluxy/onheel.htm. This is included in a
    website devoted to the entire issue: http://paleo.cc/paluxy.htm.

7   For non-UK readers: Lyme Regis is a sea-side resort in Dorset on England's South
    Coast, renowned for the sheer volume and variety of fossils to be found on its beach,
    regularly replenished by cliff-falls. Robin Hood's Bay on the Yorkshire coast has a
    similar reputation.

Appendix B

1   S.H. Hooke (1963) *Middle Eastern Mythology*. A recent, if controversial, work on the
    issue of the authorship of the Bible should be noted: Preston Kavanagh (2011) *The
    Shaphan Group. The Fifteen Authors Who Shaped the Hebrew Bible*.

Appendix C

1   The sense of *logos* as used here should not be confused with its use to refer to the
    Divine '*Logos*' or 'Divine Word'. Its meaning here is, as will become clear, closely
    related to 'logic'.

References

Aiton, E.J. (1972) *The Vortex Theory of Planetary Motions*, London: Macdonald.

Annas, G. J. (2006) 'Intelligent judging – evolution in the classroom and courtroom', *New England Journal of Medicine*, 354 (21) 2277–81.

Appelbaum, S. (2006) '"Border control"' – interview with Tommy Lee Jones', *The Big Issue*, No. 686, 27 March–2 April 2006, p.15.

Aristotle trans. J.H. McMahon (2007) *Metaphysics*, New York: Dover.

Armstrong, K. (1993) *A History of God From Abraham to the Present: The 4000-year Quest for God*, London: William Heinemann.

Armstrong, K. (2000) *The Battle for God. Fundamentalism in Christianity, Judaism and Islam*, New York: Random House.

Armstrong, K. (2006) 'Bush's fondness for fundamentalism is courting disaster at home and abroad', *The Guardian* 31 July, p.24.

Armstrong, K. (2007) *The Bible: The Biography*, New York: Atlantic Monthly Press.

Austin, J. (1962) *How to Do Things with Words. The William James Lectures Delivered at Harvard University 1955*, Oxford: Oxford University Press.

Baxter, M.P. (1887, 1st edition 1866) *Forty Coming Wonders Between 1890 and 1901 as Foreshown in the Prophecies of Daniel and Revelation*, London: Christian Herald.

Baxter, M.P. and N. Wiseman (1923, 15th edition) *Forty Future Wonders of Scripture Prophecy with a Memoir of the Author by Nathaniel Wiseman*, London: Christian Herald and Chas. J. Thynne & Jarvis.

Behe, B. (1996) *Darwin's Black Box, The Biochemical Challenge to Evolution*, Old Tappan NJ: Free Press.

Bell, C. (1832) *The Hand: Its Mechanism and Vital Endowments as Evincing Design*, London: William Pickering.

Blunt, J.H. (1874) *Dictionary of Sects, Heresies, Ecclesiastical Parties and Schools of Religious Thought*, London: Rivingtons.

Bowler, P. (1983) *The Eclipse Of Darwinism. Anti-Darwinian Evolution Theories in the Decades around 1900*, Baltimore and London: Johns Hopkins University Press.

Bowler, P. (1988) *The Non-Darwinian Revolution. Reinterpreting a Historical Myth*, Baltimore and London: Johns Hopkins University Press.

Boyce Gibson, A. (1970) *Theism and Empiricism*, London: SCM.

Broad, C.D. (1925) *The Mind and Its Place in Nature*, London: Routledge & Kegan Paul.

Cantor, G. (1991) *Michael Faraday: Scientist and Sandemanian*, New York: St. Martin's Press.

Capra, F. (1972) *The Tao of Physics. An Exploration of the Parallels between Modern Physics and Eastern Mysticism*, London: Wildwood House.

Capron, F. H. (1902) *The Conflict of Truth*, London: Hodder & Stoughton.

Carpenter, W.B. (1888) *Nature and Man. Essays Scientific and Philosophical*, London: Kegan, Trench and Trübner.

Chardin, T. de (1959) *The Phenomenon of Man*, London: Collins.

Coates, P. (1998) *Nature. Western Attitudes since Ancient Times*, Berkeley: University of California Press.

Comfort, N.C. (ed.) (2007) *The Panda's Black Box: Opening up the Intelligent Design Controversy*, Baltimore: Johns Hopkins University Press.

Cuif, J.-P., Y. Dauphin, and J. E. Sorauf (2010) *Biominerals and Fossils through Time*, Cambridge: Cambridge University Press.

Darwin, C. (1859) *On The Origin of Species by Means of Natural Selection*, London: John Murray.

Darwin, C. (1871) *The Descent of Man and Selection in Relation to Sex*, London: John Murray.

Dawes Hicks, G. (1937) *The Philosophical Bases of Theism*, London: Allen & Unwin.

Dawkins, R. (1988) *The Blind Watch-Maker. Why the Evidence of Evolution Reveals a Universe without Design*, London: Longman.

Dawkins, R. (2006) *The God Delusion*, London: Bantam Press.

Dembski, W. (ed.) (1998) *Mere Creation: Faith, Science and Intelligent Design*, Westmont IL: Intervarsity Press.

Dembski, W. (2001) *No Free Lunch: Why Specified Complexity cannot be Purchased without Intelligence*, Lanham MD: Rowan & Littlefield.

Dixon, T. (2008) *Science and Religion: A Very Short Introduction*, Oxford: Oxford University Press.

Draper, J.W. (1874) *History of the Conflict between Science and Religion*, New York: D. Appleton.

Elfe Turne, W. (1855) *Geology: Its Facts and Its Fictions*, London: Houlston & Stoneman.

Feyerabend, P. (1975) *Against Method: Outline of an Anarchistic Theory of Knowledge*, London: NLB.

Fitch, W.N. (2012) *The Three Failures of Creationism. Logic, Rhetoric, and Science*, Berkeley: University of California Press.

Freud, S. (1919) 'On the Uncanny' in J. Strachey (ed.) (1955) *Complete Psychological Works of Sigmund Freud*, Volume XVII (1917–1919), London: Hogarth Press and Institute of Psychoanalysis.

Fuller, S. (2008) *Dissent over Descent. Intelligent Design's Challenge to Darwinism*, Thriplow, Cambridge: Icon Books.

Garwood, C. (2007) *Flat Earth. The History of an Infamous Idea*, London: Macmillan.

Goodwin, B. (1994) *How the Leopard Changed its Spots: The Evolution of Complexity*, London: Weidenfeld & Nicolson.

Gould, S. J. (2000) 'More things in heaven and earth' in H. and S. Rose, pp. 85–105.

Gould, S. J. (2002) *The Structure of Evolutionary Theory*, Cambridge MA: Belknap Press.

Gould, S. J. and R. C. Lewontin (1979) 'The Spandrels of San Marco and the Panglossian Paradigm: a critique of the Adaptationist Programme', *Proceedings of the Royal Society of London, Series B.*, 205, pp. 581–98.

Haber, F.C. (1959) *The Age of the World: Moses to Darwin*, Baltimore: The Johns Hopkins University Press; Berkeley: University of California Press.

Haller, J.S. Jr (1971) *Outcasts from Evolution: Scientific Attitudes of Racial Inferiority 1859–1900*, Urbana: University of Illinois Press.

Hardy, A. (1965, 1966) *The Living Stream,* London: Collins.

Hardy, A. (1966) *The Divine Flame*, London: Collins.

Hatch, N. (1989) *The Democratization of American Christianity*, New Haven: Yale University Press.

Hooke, S.H. (1963) *Middle Eastern Mythology*, Harmondsworth: Penguin.

Hume, D. (1779) *Dialogues Concerning Natural Religion*, London: no publisher.

Jeans, J. (1930) *The Mysterious Universe*, Cambridge: Cambridge University Press.

Johnson, P.E. (1991) *Darwin on Trial*, Westmont IL: Intervarsity Press.

Jung, C.G. and W. Pauli (1955) *The Interpretation of Nature and the Psyche: Synchronicity, an Acausal Connecting Principle*, London: Routledge & Kegan Paul.

Kavanagh, P. (2011) *The Shaphan Group. The Fifteen Authors who Shaped the Hebrew Bible*, Eugene, Oregon: Wipf and Stock Publishers.

Kevles, D. J. (1985) *In the Name of Eugenics*, Cambridge MA: Harvard University Press.

Kidd, J. (1833) *On the Adaptation of External Nature to the Physical Condition of Man*, London: William Pickering.

Kinns, S. (1882) *Moses and Geology or The Harmony of the Bible with Scripture*, New York: Cassell, Petter, Galpin.

Krogman, W.M. (1951) 'Scars of evolution', *Scientific American*, 185 (6) pp.54–7.

Kugelmann, R. (2011) *Psychology and Catholicism: Contested Boundaries*, Cambridge: Cambridge University Press.

Lamont, P. (2013) *Extraordinary Beliefs. A Historical Approach to a Psychological Problem*, Cambridge: Cambridge University Press.

Larson, E. J. (1987) *Trial and Error: The American Controversy over Creation and Evolution*, New York: University Press.

Larson, E. J. (1998) *Summer for the Gods: The Scopes Trial and America's Continuing Debate over Science and Religion*, Cambridge, MA: Harvard University Press.

Latour, B. (1987) *Science in Action. How to Follow Scientists and Engineers through Society*, Milton Keynes: Open University Press.

Lawrence, J. and R. Lee (1955) *Inherit the Wind*, New York: Random House.

Lawton, G. (2009) 'Uprooting Darwin's tree', *New Scientist*, 24th January.

Lévy-Bruhl, C. (1923) *Primitive Mentality*, London: Allen & Unwin.

Lewin, R. (1993) *Complexity. Life at the Edge of Chaos*, London: J.M. Dent & Sons.

Lovejoy, C.O. (1981) 'The Origin of Man', *Science*, 211, pp.341–50.

Lovelock, J.E. (1979) *Gaia: A New Look at Life on Earth*, Oxford: Oxford University Press.

MacFadden, B. J. (1992) *Fossil Horses. Systematics, Paleobiology and Evolution of the Family Equidae*, Cambridge: Cambridge University Press.

Mangum, R.T. and M.S. Sweetnam (2009) *The Scofield Bible: Its History and Impact on the Evangelical Church*, Colorado Springs: Paternoster.

Martineau, J. (1875) *Religion as Affected by Modern Materialism. An Address delivered in Manchester New College, London*, New York: Putnams.

Martineau, J. (1890) *The Seat of Authority in Religion*, London: Longmans, Green.

McReady Price, G. M. (1906) *Illogical Geology*, Mountain View, CA: Pacific Press Publishing Association.

McReady Price, G. M. (1913) *The Fundamentals of Geology*, Mountain View, CA: Pacific Press Publishing Association.

McReady Price, G. M. (1923) *The New Geology*, Mountain View, CA: Pacific Press Publishing Association.

McReady Price, G. M. (1926) *Evolutionary Geology and the New Catastrophism*, Mountain View, CA: Pacific Press Publishing Association.

Meyer, S. C. (2002) 'The Scientific status of Intelligent Design: the methodological equivalence of naturalistic and non-naturalistic origins theories' in Michael J. Behe, William A. Dembski, Stephen C. Meyer (eds) *Science and Evidence for Design in the Universe*, San Francisco: Ignatius Press, pp.151–211.

Midgley, M. (1985) *Evolution as a Religion*, London: Methuen.

Midgley, M. (1992) *Science as Salvation: A Modern Myth and Its Meaning*, London: Routledge.

Midgley, M. (2007) *Intelligent Design Theory and Other Ideological Problems*, London: Philosophy of Education Society of Great Britain, Impact 15.

Mooney, C. (2005) *The Republican War on Science*, New York: Basic Books.

Nelkin, D. (1982) *The Creation Controversy. Science or Scripture in the Schools*, New York: Norton.

Nelkin, D. (2000) 'Less selfish than sacred? Genes and the religious impulse in evolutionary psychology' in H. and S. Rose (eds), pp.14–27.

Newell, N.D. (1956) 'Fossil populations' in P.C. Sylvester-Bradley (ed.) *The Species Concept in Palaeontology*, London: The Systematics Association, Publication No.2, pp. 63–82.

Noll, M. (2002) *America's God: from Jonathan Edwards to Abraham Lincoln*, Oxford: Oxford University Press.

Numbers, R. L. (1987) 'The Creationists', *Zygon* 22 (2) pp.133–64.

Numbers, R. L. (2006) *The Creationists: The Evolution of Scientific Creationism*, Cambridge MA.: Harvard University Press.

Otto, R. (1923) *The Idea of the Holy*, Oxford: Oxford University Press.

Paley, W. (1800) *Natural Theology*, London: R. Faulder.

Pennock, R.T. (1999) *Tower of Babel. The Evidence against the New Creationism*, Cambridge MA and London: MIT Press.

Pennock, R.T. (ed.) (2001) *Intelligent Design, Creationism and Its Critics: Philosophical, Theological, and Scientific Perspectives*, Cambridge MA: MIT Press.

Pick, D. (1989) *Faces of Degeneration. A European Disorder c.1848–c.1919*, Cambridge: Cambridge University Press.

Pickering, T.R. (ed.) (2007) *Breathing Life into Fossils: Taphonomic Studies in Honor of C.K. ('Bob') Brain*, Bloomington IN: Stone Age Institute Press.

Popper, K. (1959) *The Logic of Scientific Discovery*, London: Hutchinson.

Raleigh, Sir W. (1614, 4th ed. 1628) *The Historie of the World*, London: H. Lownes, G. Lathum, and R. Young.

Raven, C. E. (1942) *John Ray, Naturalist. His Life and Works*, Cambridge: Cambridge University Press.

Ray, J. (1691, 2nd ed. 1692) *The Wisdom of God Manifested in the Works of Creation*, London: Samuel Smith.

Rendell, E.D. (1864 2nd ed., 1st ed.1850) *The Antediluvian History, and Narrative of the Flood, as Set Forth in the Early Portions of the Book of Genesis. Critically Examined and Explained*, London: F. Pitman.

Richards, G. (1987) *Human Evolution: An Introduction for the Behavioural Sciences*, London: Routledge & Kegan Paul.

Richards, G. (1989) *On Psychological Language and the Physiomorphic Basis of Human Nature*, London: Routledge.

Richards, G. (2005) 'The psychology of explanation', *History and Philosophy of Psychology* 7(1), 53–61.

Richards, G. (2010 3rd ed.) *Putting Psychology in its Place. A Critical Historical Overview*, London: Psychology Press.

Richards, G. (2011) *Psychology, Religion, and the Nature of the Soul. A Historical Entanglement*, New York: Plenum Press.

Richards, G. (2012, 2nd ed., 1st 1997) *'Race', Racism and Psychology. Towards a Reflexive History*, London: Routledge.

Richards, G. and S. Fuller (2009) 'Review of Steven Fuller Dissent over Darwin' with Fuller's response and reviewer's reply, *History of the Human Sciences* 22 (5), 113–26.

Rose, H. and S. Rose (eds) (2000) *Alas, Poor Darwin. Arguments Against Evolutionary Psychology*, London: Jonathan Cape.

Rowbotham, S. B. as 'Parallax' (1865) *Zetetic Astronomy. The Earth not a Globe!*, London: Simpkin Marshall.

Ruse, M. (2006) *Darwinism and Its Discontents*, Cambridge: Cambridge University Press.

Ruse, M. (ed.) (1988) *But Is It Science? The Philosophical Question in the Creation/ Evolution Controversy*, Buffalo: Prometheus Books.

Sarkar, S. (2007) *Doubting Darwin? Creationist Designs on Evolution*, Oxford: Blackwell.

Schrödinger, E. (1958) *Mind and Matter. The Tarner Lecture 1956*, Cambridge, Cambridge University Press.

Scofield, C.I. (ed.) (1909) *The Scofield Reference Bible*, New York: Oxford University Press.

Shipman, P. (1981) *Life History of a Fossil and Introduction to Paleoecology*, Cambridge MA: Harvard University Press.

Simon, R. (Paris 1678, rep.1685) *Histoire Critique Du Vieux Testament*, Rotterdam: Reinier Leers.

Smith, A. (1776) *The Wealth of Nations*, London: Strahan & Cadell.

Smith, R. (2013a) *Between Mind and Nature. A History of Psychology*, London: Reaktion Books.

Smith, R. (2013b) *Free Will and the Human Sciences in Britain, 1870–1910*, London: Pickering & Chatto.

Sprat, T. (1667) *History of the Royal Society*, London: J. Martyn.

Tillich, P. (1952) *The Courage to Be*, New Haven: Yale University Press.

Tucker, W. (1994) *The Science and Politics of Racial Research*, Urbana and Chicago: University of Chicago Press.

Ussher, J., J. Flesher, and L. Sadler (1650) 'Annales Veteris Testamenti, a prima mundi origine deducti' in R.E. Elrington (ed.) *The Whole Works of the Most Rev. James Ussher D.D.* Volume VIII, Dublin: Hodges & Smith.

Weatherhead, Leslie D. (1965) *The Christian Agnostic*, London: Hodder & Stoughton.

Wells, H. G. (1904, reprinted 1911) 'The Country of the Blind', *The Strand Magazine*, April; reprinted in *The Country of the Blind and Other Stories*, London: Thomas Nelson.

Whewell, W. (1833) Astronomy and General Physics Considered with Reference to Natural Theology, London: William Pickering.

Whitcomb, J.C. Jr and H.M. Morris (1960) *The Genesis Flood. The Biblical Record and Its Scientific Implications*, Phillipsburg, NJ: P.& R. Publishing.

Zimmer, H. (ed. J. Campbell) (1946) *Myths and Symbols in Indian Art and Civilization*, Princeton: Princeton University Press.

Index

www.ingramcontent.com/pod-product-compliance
Lightning Source LLC
Chambersburg PA
CBHW020858090426
42736CB00008B/417